High-Resolution Electronic Spectroscopy of Small Molecules

High-Resolution Electronic Spectroscopy of Small Molecules

Geoffrey Duxbury
Alexander Alijah

CRC Press
Taylor & Francis Group
Boca Raton London New York

CRC Press is an imprint of the
Taylor & Francis Group, an **informa** business

CRC Press
Taylor & Francis Group
6000 Broken Sound Parkway NW, Suite 300
Boca Raton, FL 33487-2742

First issued in paperback 2020

ISBN 13: 978-0-367-57376-8 (pbk)
ISBN 13: 978-1-4822-4559-2 (hbk)

Visit the Taylor & Francis Web site at
http://www.taylorandfrancis.com

and the CRC Press Web site at
http://www.crcpress.com

Contents

CHAPTER 1

Introduction

CONTENTS

1.1 HISTORICAL DEVELOPMENT OF ELECTRONIC SPECTROSCOPY

Electronic spectroscopy was one of the first techniques used to study the absorption and emission spectra of small molecules, as photographic plates could be used to record electronic spectra in the visible and ultraviolet regions. At longer wavelengths corresponding to the infrared and microwave regions, suitable detectors were only developed in the period from 1938 onwards in response to military requirements. The first detailed review of the "Electronic Spectra of Polyatomic Molecules" was given by H. Sponer and E. Teller in the April 1941 issue of *Reviews of Modern Physics* [1].

In the period following the end of the Second World War, some of the instruments that had been developed in the period from 1938 to 1946 began to be used for spectroscopic purposes. One of the first of these methods, flash photolysis, was developed by Norrish and Porter at Cambridge University [2]. The 1967 Nobel Prize in Chemistry was divided, one half awarded to Manfred Eigen, the other half jointly to Ronald George Wreyford Norrish and George Porter "for their studies of extremely fast chemical reactions, effected by disturbing the equilibrium by means of very short pulses of energy."

In 1950, George Porter [3] described the development of "Flash Photolysis and Spectroscopy. A New Method for the study of Free Radical Reactions." This formed part of Porter's Nobel lecture in December 1967 [4]. Flash photolysis required the use of "gas-filled discharge tubes of very high power." Their use allowed measurements of an absorption spectrum of a gas phase intermediate in "one twenty thousandth of a second in short intervals afterwards." Some of the species detected were free radicals.

This was followed by the development of "Nanosecond Flash Photolysis" by Porter and Topp in 1970 [5,6]. This allowed the detection of events having a nanosecond (ns) timescale, and led to a switch from the use of photographic plates to photoelectric monitoring. Much later, Topp and his group developed a femtosecond up conversion spectrometer [7].

A systematic study of gas phase free radicals was undertaken by Herzberg and Ramsay in 1952. They were the first to identify "Free NH_2 radicals" in absorption using flash photolysis, and in emission from an ammonia oxygen flame [8]. Later, in 1957, K. Dressler and D. A. Ramsay described the splitting of the $^2\Pi$ ground state of NH_2 into a $X\,^2B_1$ ground state and an $A\,^2A_1$ excited state, with the electronic transitions occurring between these two half states. This was one of the first examples of the "Renner, or Renner–Teller, Effect" [9].

In 1958 J. A. Pople and H. C. Longuet-Higgins developed a "Theory of the Renner Effect in the NH_2 radical." This was published in the first edition of *Molecular Physics* [10]. A more complete analysis of the Renner–Teller effect in the spectra of NH_2 and ND_2 was given by K. Dressler and D. A. Ramsay in *Philosophical Transactions of the Royal Society of London* [11].

In 1960, R. N. Dixon analysed the absorption spectrum of the free NCO radical, and showed that it was due to an electronic transition, the $A\,^2\Sigma^+ \leftarrow X\,^2\Pi_i$, an example of the Renner effect in a linear molecule [12]. This was followed in 1960 by the analysis of the second example of the Renner effect, the $B\,^2\Pi \leftarrow X\,^2\Pi_i$ transition of NCO [13].

The other free radical which was analysed in detail by Herzberg [14], and by Herzberg and Johns [15], was methylene, CH_2. They showed that the ground state was a triplet state $(^3\Sigma_g^-)$ [14] and that there is a low-lying singlet state, $a\,^1A_1$ in which the molecule is well bent, and an excited $b\,^1B_1$ state, which is almost linear at equilibrium, the two states forming the two components of a Renner–Teller pair [15].

1.2 HIGH-RESOLUTION LASER SPECTROSCOPY

After G. Duxbury, Chem. Soc. Rev. **12**, 453–504 (1983) with permission from The Royal Society of Chemistry.

1.2.1 Introduction

The rapid development of lasers since their first use as spectroscopic tools in 1964 has resulted in very-high-resolution spectrometers being available for the wavelength region from the infrared to the ultraviolet region, i.e., from 1 mm to 300 nm. The availability of laser sources has completely improved the sensitivity, resolution and selectivity, which are still improving.

In this article, we will consider both the armory of techniques now available for carrying out high-resolution spectroscopy, and also the use of combinations of methods for studying small polyatomic molecules, which are of considerable interest because of chemical reactions, or for testing current models of vibration-rotation interaction and of dipole moment variation.

1.2.2 Principles of Sub-Doppler Spectroscopy

In order to appreciate the reasons for the high resolution and the high sensitivity of laser spectroscopy, it is first of all necessary to consider the details of the interaction between monochromatic high-power coherent sources and molecular absorbers. This comprises both the mechanisms of spectral line broadening in low-pressure gases, and also the perturbations of the line shape associated with the strong laser radiation field. The various types of available spectrometers can then be classified not only from the point of view of the resolution and the selectivity available, but also from the standpoint of analogies between laser spectrometers and magnetic resonance or microwave spectrometers currently in use in university chemistry or physics departments.

Line broadening in low-pressure gases

Lines of sub-Doppler spectroscopy in low-pressure gases can be classified into two types, homogeneous and heterogeneous. The general form of a homogeneous broadened line is Lorenzian, with the absorption coefficient given by

$$\alpha(\nu) = \frac{S_1}{\pi} \frac{\gamma}{(\nu - \nu_0)^2 + \Gamma^2}, \tag{1.1}$$

where Γ, the power broadened line width, is given by

$$\Gamma^2 = \gamma^2 + \frac{|\mu_{12}|^2 E^2}{h^2}. \tag{1.2}$$

γ, the half width at half height in the absence of saturation, is the sum of contributions from natural line width, pressure broadening, and transit time effects. μ_{12} is the transition matrix element between levels 1 and 2, which we abbreviate as μ, E is the electric field amplitude of the incident radiation field, and S_1 is the integrated absorption coefficient for the line.

Heterogeneous line broadening in gases is associated with the Doppler effect. If a moving molecule emits radiation, the emitted radiation is not centered at the resonant frequency of the molecular transition, but is shifted by an amount that depends on the molecular velocity in the direction of the emitted radiation. The Doppler effect occurs because the moving wave fronts of the emitted radiation are compressed in the direction of the molecule's motion, and are expanded in the opposite direction, giving rise to a shift in wavelength and, hence, its frequency. The apparent emission frequency is increased if the molecule moves towards the observer, and is decreased if the molecule moves away. A similar effect occurs if we consider the absorption of radiation by a molecule. If we choose the $+z$-direction in the laboratory to coincide with the direction of propagation of a laser beam, the frequency ν_1 of a laser in the laboratory frame of reference appears in the frame of reference moving with the molecule as

$$\nu' = \nu_1 \left(1 + \frac{\nu_z}{c}\right), \tag{1.3}$$

where ν_z is the component of the molecular velocity along the z-direction in the laboratory frame. The molecule can only absorb radiation if ν' coincides with its absorption frequency ν_0, i.e., $\nu_1 = \nu_0 \left(1 + \nu_z/c\right)$. The

$$w(v_z) = \frac{1}{u\sqrt{\pi}} e^{-\frac{\nu_z^2}{u^2}} \tag{1.4}$$

absorption frequency of the laser radiation is, thus, higher than the centre frequency if ν_z is positive, i.e., if the molecule moves parallel to the wave propagation direction, and is lower than the centre frequency if ν_z is negative, i.e., the molecule moves in the opposite sense to the direction in which the laser beam is propagating. These effects are shown schematically in Figure 1.1.

The components of the molecular velocities along the fixed direction in a gas at thermal equilibrium at temperature T obey the Maxwell distribution,

$$w(v_z) = \frac{1}{u\sqrt{\pi}} e^{-\frac{\nu_z^2}{u^2}}, \tag{1.5}$$

where $u^2 = 2kT/m$, m is the molecular weight, and k is Boltzmann's constant. This gives rise to the Gaussian lineshape

$$a(\nu) = \frac{S_1}{\delta\pi} e^{-\left(\frac{\nu-\nu_0}{\delta}\right)^2} = a_0 e^{-\left(\frac{\nu-\nu_0}{\delta}\right)^2}, \tag{1.6}$$

where the half width parameter δ is given by

$$\delta = \frac{v}{c}\sqrt{\frac{2kT}{m}}. \tag{1.7}$$

The half width at half height is related to $\delta(\ln 2)^{1/2}$, and at maximum slope by $\delta/\sqrt{2}$.

Line shapes in gases are governed by both the homogeneous and the heterogeneous contributions. In the infrared region the line shape may be considered to be made up from a large number of Lorentzian curves, one for each group of molecular velocities. This is shown schematically in Figure 1.1. The resultant line shape is a convolution of the Gaussian and the Lorentzian contributions and is known as the Voigt line shape. In the spectra

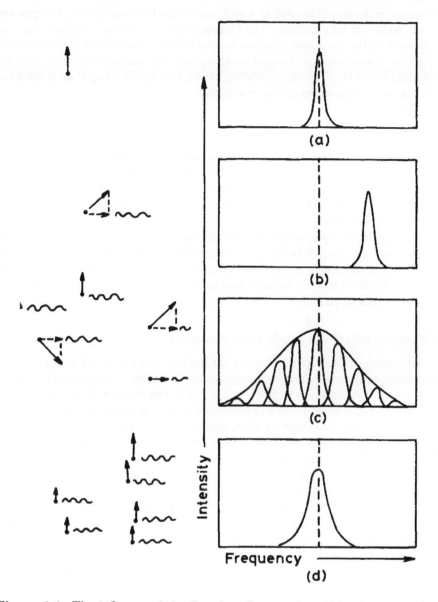

Figure 1.1: The influence of the Doppler effect on the width of a spectra line. (a) Homogeneous (Lorentzian) line shape, centred at ν_0, if the particle is moving perpendicular to the laser beam direction, z. (b) Doppler-shifted line position for one velocity component along z. (c) Envelope of the line shapes of (a) and (b) produced when particles can move in all directions. (d) In a molecular beam, the spread of velocities along the z-direction is small, and area (a) is almost recovered. Reproduced from Duxbury, *Chem. Soc. Rev.*, 12:453, 1983 [16], with permission from The Royal Society of Chemistry.

of many molecules observed at wavelengths shorter than $20\,\mu$m, the predominant source of broadening is the Doppler effect, with the homogeneous contribution to the linewidth being at least a factor of one hundred times smaller than the heterogeneous Doppler contribution. The partial or almost complete elimination of the Doppler contribution to line broadening thus results in a very great enhancement of the resolution.

The oldest form of sub-Doppler spectroscopy is atomic or molecular beam spectroscopy, in which a collimated beam of atoms or molecules is directed at right angles to the radiation field. The reduction in Doppler broadening obtainable is then dependent on the spread of the beam. The main problems encountered are associated with the low pressures in molecular beam spectrometers, which limits the sensitivity of the system, and the difficulties inherent in producing intense, well-collimated beams. The more recent methods of sub-Doppler spectroscopy make use of the properties of *the radiation field itself* to pick out a particular velocity group from the random set of molecules in the bulk gas sample in the cell. This results in a higher sensitivity than is generally in molecular beam experiments; it also results in much higher resolution than can usually be obtained using molecular beam spectrometers in the infrared and the visible spectral regions.

Saturation effects and sub-Doppler linewidths

If we consider the effect of a strong saturating field produced by a laser, on a predominantly heterogeneously broadened molecular absorption line, we can see from the discussions in Section 1.2.2 that the effect will be confined to a narrow region centred upon the velocity component of the group of molecules whose absorption is Doppler shifted into resonance. This produces a "hole" in the population of the lower state and a "spike" in the population of the upper state, resulting in a hole in the absorption profile. This "hole burning" is due to Bennett [17,18], and is shown schematically in Figure 1.2 for interaction of molecules with a running wave propagating in the $+z$-direction.

Since the molecular velocity component at resonance will have a homogeneous lineshape, the "hole," which depends upon the partial saturation of the Lorenzian line centred at ν_z, will also have a Lorenzian shape. Thus, the hole can frequently be treated as an absorption feature with a negative absorption coefficient. The width of the hole is determined by the factors governing homogeneous line broadening, i.e.,

$$\gamma_{hole} = \left[\gamma^2 + \left(\frac{\mu E}{h}\right)\right]^{\frac{1}{2}} = \Gamma. \tag{1.8}$$

The various methods of saturation spectroscopy rely on detecting this narrow hole by differential saturation effects, and, hence, achieving considerable resolution enhancement.

where the collisions involve very little change in molecular velocities, four-level resonances may be observed, as first described by Shoemaker et al. [23] and Johns et al. [24]. This situation is shown schematically in Figure 1.3. The frequency-matching condition is identical to that for three-level resonances, but the intensities are different, since not all collisions are effective in coupling the levels within the four-level system without change in velocity.

Optical-optical double resonance

If a pair of transitions sharing a common level and overlapping to within about two Doppler widths interact with two copropagating laser beams of different frequencies, a nonlinear response will occur for the resonance condition [25,26]

$$\Omega_1 - \Omega_2 = \nu_1 - \nu_2 \tag{1.10}$$

or

$$\Omega_1 - \nu_1 = \Omega_2 - \nu_2 = \frac{v_z \bar{\nu}}{c}, \tag{1.11}$$

where $\Omega_1 - \Omega_2$ is the difference frequency between the lasers, $\nu_1 - \nu_2$ is the difference frequency between the molecular transitions, and $\bar{\nu} = (\nu_1 - \nu_2)/2$. These experiments are usually carried out by keeping $\Omega_1 - \Omega_2$ fixed, and by varying $\nu_1 - \nu_2$ by the application of either an electric or magnetic field. Since the signal depends on a particular velocity group, v_z, being saturated, the linewidths are similar to those observed for Lamb dips.

Anti-crossing

If avoided crossings between Zeeman [27] or Stark [28] sublevels occur, sub-Doppler spectra can be observed that resemble those obtained in level-crossing experiments.

Doppler-free two-photon signals

If we consider the interaction of a molecule with velocity v_z, interacting with a standing wave field of frequency ν_1 in the frame of reference of the molecule, the molecule interacts with two oppositely travelling waves of frequencies $\nu_1(1 - v_z/c)$ and $\nu_1(1 + v_z/c)$, respectively. If the molecule can be excited from the ground to the excited state by the absorption of two photons, one from each of the oppositely directed travelling waves, then at resonance the following condition is fulfilled:

$$\frac{E' - E''}{h} = \nu_1\left(1 - \frac{v_z}{c}\right) + \nu_1\left(1 + \frac{v_z}{c}\right) = 2\nu_1, \tag{1.12}$$

i.e., the resonance is independent of the velocity, v_z, [29–31]. In practice, two-photon transitions will be introduced by the absorption of photons from travelling waves that are copropagating, as well as the signal described above

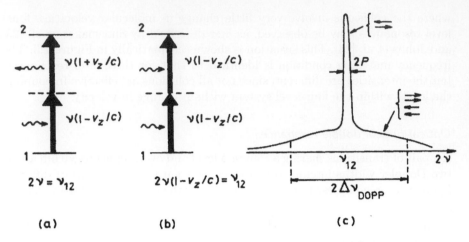

Figure 1.4: (a) Compensation for the Doppler shift in the simultaneous absorption of two photons from two counter-running waves. (b) Absence of compensation from two unidirectional waves (c) Shape of the narrow resonance in two-photon absorption. Reproduced from Duxbury, *Chem. Soc. Rev.*, 12:453, 1983 [16], with permission from The Royal Society of Chemistry.

that results from the absorption of the two photons from counter-propagating waves. The Doppler-free, two-photon signal will, therefore, be superimposed upon a broad Doppler broadened base, as shown in Figure 1.4. A fundamental difference exists between the narrow signals seen in two-photon experiments of this type and the narrow signals described in the previous sections. The two-photon signals arise from *all* molecules that observe one photon from each of the counter-propagating beams, whereas the narrow signals described in the previous sections depend on the contribution of only one particular velocity component of the group of molecules to the signal.

A final way of reducing the Doppler broadening in the infrared and visible regions is to shift the detection from the optical to the microwave region, where the Doppler widths are very small [32–34]. These double resonance methods involve either saturating the optical transition and detecting the effects of this on the microwave absorption, or vice versa. The observation of double resonance signals requires either that the optical and microwave transitions share a common level, or that the levels are strongly coupled by collisions that obey well-defined selection rules.

1.2.3 Experimental Techniques

Resonance techniques

Resonance methods that rely on "tuning a molecule" rather than a laser were the first techniques to be developed for Doppler-limited and saturation spectroscopy, since lasers with a good broadband tuning range were not available until the early 1970s. The first two methods to be developed relied on the use of electric or magnetic fields to tune a particular set of transitions into resonance with a stable fixed-frequency gas laser. These methods are known as "laser Stark spectroscopy" in the case of electric field tuning [35–40] and "laser magnetic resonance" (LMR) when magnetic field tuning is used [41–44].

In laser Stark spectroscopy, fixed-frequency laser radiation is passed through a parallel plate Stark cell. At certain values of the applied electrostatic field-specific electric-field components of the molecular vibration-rotation, or pure rotation, transitions of the absorbing gas are brought into resonance with the laser frequency, giving rise to an electric resonance spectrum. In the infrared region, the short wavelength radiation propagates through the Stark cell as a free-space wave, and, hence, the electric vector of the radiation field can be both parallel and perpendicular to the static electric field, leading to both $\Delta M_J = 0$ and $\Delta M_J = \pm 1$ selection rules in the observed spectra.

Since the method relies on the use of electrostatic tuning, it is necessary to generate high uniform-fields and to study molecules with appreciable Stark type tuning coefficients. In order to generate high electrostatic-fields, which may approach 90 kV cm^{-1}, narrow plate spacings from 1 to 4 mm are commonly used. With such narrow gaps, the plates must be flat to within one or two fringes of visible light, and be held accurately parallel. The gas pressure used must also be restricted to the low-pressure region below 100 millitorr. A useful rule of thumb is that a potential difference of 3000 volts may be sustained without electrical breakdown across any gas at a pressure of 10 millitorr and below.

The detectors used are quantum limited liquid nitrogen cooled semiconductor devices, PbSnTe or CdHgTe in the 10 μm region, and Au doped Ge in the 5 μm region. Modulation frequencies of 5 kHz to 100 kHz are used to get above the principal noise region of 5 μm to 10 μm gas lasers.

Intracavity [45] or multiple-pass absorption cells of the Shimizu type [46] are generally used. In both types of cells, Lamb dips can usually be seen due to the presence of a standing-wave field. The main disadvantage of the intracavity cells is that the spot size of most infrared gas lasers restrict their use to the 5 μm region, so that only low fields can be achieved, as the plate spacings must be ca. 3 mm to 4 mm.

The transitions seen in the ν_3 band of CH_3F form a useful vehicle for discussion of typical laser Stark patterns [47]. The 9 μm $P(18)$ line of the CO_2 laser lies very close to the ν_3 vibrational band origin. In perpendicular polarization, the lowest J ^QQ transitions are seen with $\Delta M_J = \pm 1$ selection rules, as shown in Figure 1.5. The spectrum consists of a set of ^QQ transitions $(^{\Delta K}\Delta_J)$ that are brought into resonance at differing values of the static electric field. The first derivative shape is due to the use of small amplitude sinusoidal modulation, and is similar to that observed in electron spin resonance (ESR) spectra [48].

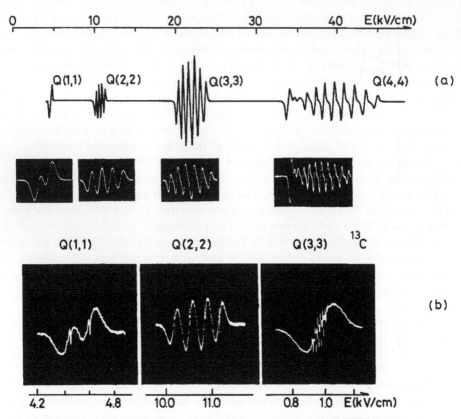

Figure 1.5: (a) Observed $\Delta_M = \pm 1$ transitions for the Q-branch series of $^{12}CH_3F$. The laser line used is the CO_2, $9\,\mu m$ $P(18)$ line. The upper trace is a computer calculated band contour, and the lower picture shows observed oscilloscope traces. The features on the shoulder belong to the $Q(3,2)$ pattern. The sample pressure was about 5 mtorr, and the time constant for detection was 10 ms. (b) $Q(1,1)$, $Q(2,2)$, and $Q(3,3)$ in $^{13}CH_3F$ transitions with Lamb dip resolution. Reprinted from Freund et al., *J. Mol. Spectrosc.* 52:38, 1974 [47]. Copyright (1974), with permission from Elsevier.

The origin of the patterns can be seen using the energy level diagram of Figure 1.6.

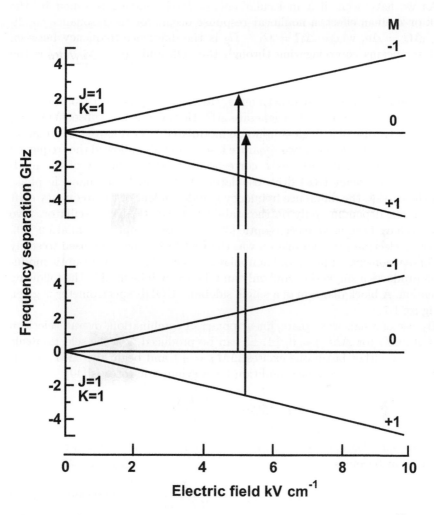

Figure 1.6: Stark energy level diagram for the $Q(1,1)$ transition of $^{12}CH_3F$ with $\Delta M = \pm 1$. Since the Stark shifts of the excited and ground states are very similar, the splitting of the $Q(1,1)$ resonance in Figure 1.5 is caused by the very small difference between the dipole moments in the upper and lower states of CH_3F. Reprinted from Freund et al., *J. Mol. Spectrosc.*, 52:38, 1974 [47]. Copyright (1974), with permission from Elsevier.

In general, the energy level patterns seen in LMR spectra are more complex than their electric field equivalents, mainly as a result of two additional complications: the interplay of the various magnetic moments associated with the unpaired electron and the nuclear spins, and the effects of spin-uncoupling at high magnetic fields, the Paschen-Back effect.

As we have seen, if a molecular energy level splitting is tuned by the Stark or Zeeman effect, a nonlinear response occurs for the resonance condition, $\Delta\Omega = \Delta\nu$, where $\Delta\Omega = \Omega_1 - \Omega_2$ is the difference frequency between the laser beams copropagating through the cell, and $\Delta\nu = \nu_1 - \nu_2$ is the difference frequency between the coupled transitions that share a common energy level. This is the principle of Optical-Optical Double Resonance (OODR) spectroscopy.In the earliest experiments, carried out by Brewer [25], two very stable CO_2 lasers were used to generate $\Delta\Omega$. However, it was subsequently realised that the use of a single-amplitude modulated laser removed the requirement for frequency locking two separate lasers, and also allowed the frequency difference to be set directly, as it corresponds to the frequency of the modulator radiofrequency (RF) drive oscillator. As the sidebands move in phase with the carrier, the difference frequency is independent of the frequency drift of the laser, depending only on the stability of the RF drive oscillator. Two methods have been used, electro-optic [39] and acousto-optic modulation [40]. Although electro-optic modulation was the first technique to be used to study the dipole moments of polar and nonpolar molecules [39], acousto-optic modulation produces a single sideband only and, hence, will be used in the following discussion. A block diagram of a single sideband OODR spectrometer is given in Figure 1.7.

By use of a half-wave plate/linear polarizer combination, overall selection rules of ΔM_J (or ΔM_F) $= 0, \pm1, \pm2$ can be produced. For the selection rule $|\Delta M_J| = 2$, either beam one excites $\Delta M_J = +1$ and beam two $\Delta M_J = -1$, or vice versa. The resonance condition for a symmetric rotor is [40]

$$\Delta\Omega = \Omega_{RF} = \frac{2\mu EK}{hJ(J+1)}.$$ (1.13)

If the overall selection rule is $|\Delta M_J| = 1$, the resonance condition for a symmetric rotor becomes

$$\Delta\Omega = \Omega_{RF} = \frac{\mu EK}{hJ(J+1)}.$$ (1.14)

When the double resonance is carried out in a multiple-pass cell, signals due to Lamb dips and three- or four-level velocity-tuned double resonances can be seen as well as OODR signals. An example of these signals in CH_3OH is shown in Figure 1.8, where the use of polarization selective detection has allowed the removal of the unwanted standing wave signals.

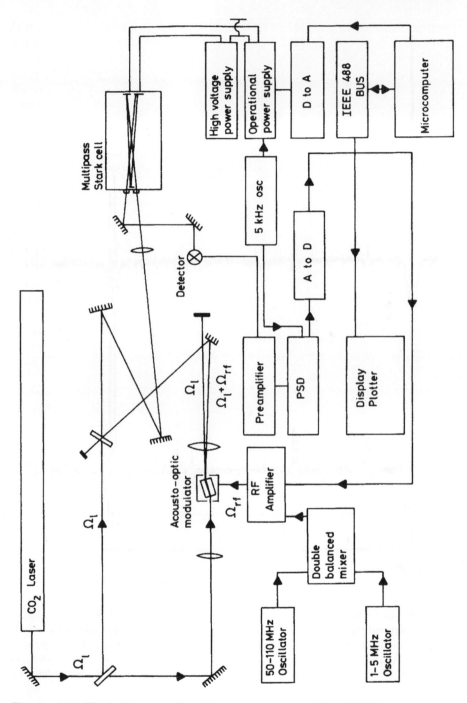

Figure 1.7: Block diagram of a microcomputer controlled OODR spectrometer. Reproduced from Duxbury, *Chem. Soc. Rev.*, 12:453, 1983 [16], with permission from The Royal Society of Chemistry.

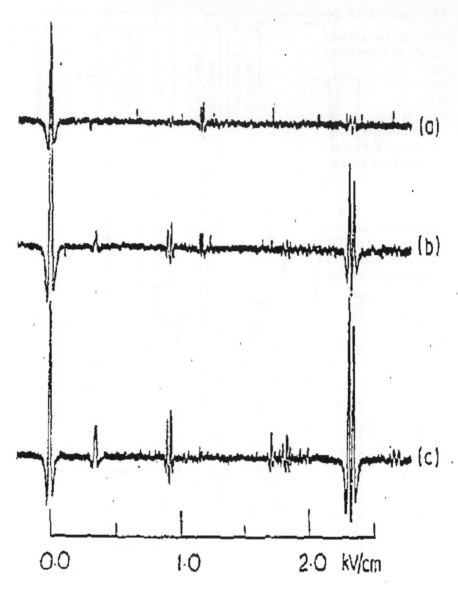

Figure 1.8: OODR spectra of CH_3OH using the $P(34)$ laser line of CO_2. (a) Two beams perpendicularly polarized, detecting in perpendicular polarisation. (b) One beam perpendicular and one parallel polarised, detecting in perpendicular polarisation. (c) As (b) but detecting in parallel polarization. Reprinted from Bedwell and Duxbury, *Chem. Phys.*, 37:445, 1979 [40]. Copyright (1983), with permission from Elsevier.

The experimental detection of sub-Doppler signals

Although the principles of sub-Doppler spectroscopy described before apply equally to experiments carried out in the infrared and the visible wavelength regions, the methods used to detect the sharp resonant signals are usually rather different. In the infrared region, the spontaneous emission probability of the excited state is usually rather low, and, hence, the saturated absorption signal is detected via the change in the absorption coefficient. In the visible region, however, the spontaneous emission probabilities are usually quite large, and, hence, the saturation of the transition is often monitored via spontaneous emission from the excited-state energy levels. In level crossing and anticrossing experiments, the differences between experiments in the two regimes are more marked. Linear-level crossing and anti crossing signals can be observed via fluorescence detection but cannot be observed in direct absorption. The signals observed via laser-induced fluorescence can, therefore, be due to mixtures of the linear and nonlinear signals, whereas in absorption only the nonlinear, or stimulated, signals can be seen.

Another distinction that can be made in the detection of signals induced by the interaction of molecules with a standing wave field is whether the standing wave field is produced by a two-beam or a multiple-beam system. Two-beam systems are frequently used in electronic spectroscopy, a typical arrangement being shown in Figure 1.9 (a). The finite crossing angle of the beams results in an incomplete removal of the Doppler broadening, but the separation of the pump and the probe beams allows various tricks to be employed to enhance the selectivity of the system.

Two of the most commonly employed are intermodulated fluorescence [49] and polarization spectroscopy [50]. In intermodulated fluorescence the pump and the probe beams are chopped at frequencies ν_1 and ν_2, the fluorescence signal at either the sum frequency $\nu_1 + \nu_2$, or the difference frequency $\nu_1 - \nu_2$ is detected. Since both the pump and the probe beams are necessary to detect the Lamb dip, only the dip is modulated at the sum or the difference frequency, the background being modulated at either ν_1 or at ν_2 is rejected by the phase-sensitive detection system. In polarization spectroscopy experiments, the pump beam is circularly polarized and the probe beam linearly polarized, as shown in Figure 1.9 (b). The circularly polarized beam induces an optical anisotropy in the sample, so that the atoms or molecules "dressed" by the circularly polarized field rotate the plane of polarization of the probe beam. The molecule interacting with a strong "handed" field thus behaves like a chiral molecule, and the rotated probe beam can then pass through the analysing polarizer, which is crossed for the original direction of linear polarization. Since only the molecules with $v_z = 0$ can be both pumped and probed in this way, only Lamb dip signals are detected. Various versions of these methods, such as polarization-intermodulated excitation (POLINEX) have subsequently been used.

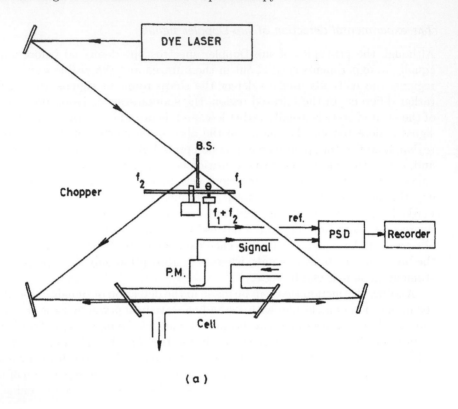

(a)

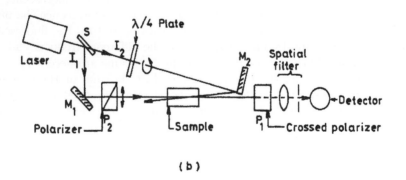

(b)

Figure 1.9: Two-beam saturation spectroscopy arrangements. (a) Intermodulated fluorescence. (b) Polarisation spectroscopy. Reprinted (a) from Sorem and Schawlow, *Opt. Commun.* 5:148, 1972 [49]. Copyright (1972), with permission from Elsevier. Reprinted (b) with permission from Wieman and Hänsch, *Phys. Rev. Lett.* 36:1170, 1976 [50]. Copyright (1976) by the American Physical Society.

In the infrared region, the principal methods used for sub-Doppler spectroscopy have involved the use of multiple-beam systems. These consist of

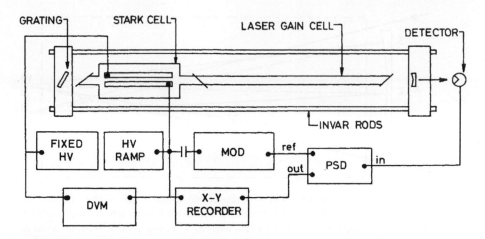

Figure 1.10: Schematic diagram of an intracavity laser Stark spectrometer. PSD stands for a phase sensitive detector, DVM for digital voltmeter, HV for high voltage, and mod for modulation source. Reproduced from Johns and McKellar, *J. Chem. Phys.* 66:1217, 1977 [45], with the permission from AIP Publishing.

either multiple-pass cells in the Shimizu configuration [46] or in intracavity cells [45]. These are shown in Figures 1.10 and 1.11. An intracavity cell can be considered as an example of a multiple-beam system, since the Fabry–Perot cavity of the laser acts as a multiple-pass system for the generation of the laser signal, and, hence, the absorber can be considered to affect the gain for each pass. In multiple-pass systems, the forward and the backward waves are equal in intensity, and selective modulation of the pump and the probe beams is not practicable. However, the detection systems commonly used do, in fact, allow the sub-Doppler signals to be enhanced relative to the background.

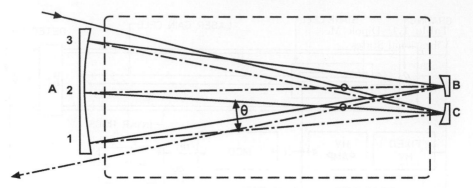

Figure 1.11: Mirror configuration of a crossover White's cell of the Shimizu type with $N = 4$. The solid lines show beams propagating from left to right, and the broken lines from right to left. The circles are the crossing points between solid and broken lines. The angle θ represents the finite crossing of the beams. The Stark electrodes are parallel to the paper and occupy the area shown by the broken line rectangle. Reprinted from Caldow et al., *J. Mol. Spectrosc.* 69:239, 1978 [46]. Copyright (1978), with permission from Elsevier.

1.2.4   Stark Spectroscopy of Small Molecules

Stark spectroscopy has been used for two main purposes: as a high-resolution spectroscopic method for measuring the absorption spectra of low pressure gases, particularly of short-lived species, and for the accurate measurement of electric dipole moments. It soon became evident that although the *absolute* precision of dipole moment determination by laser methods was little better than that achieved in the best microwave spectrometers, and was inferior to that of microwave molecular beam systems, the laser-based spectrometers provided a more accurate way of measuring *changes* in dipole moment than is possible in most microwave spectroscopy experiments. Furthermore, for some nonpolar molecules, only laser techniques were suitable for the measurement of the small vibrationally and rotationally induced dipole moments.

Dipole moments

The variation of the dipole moment of a molecule with its state is associated with the distortion of the electronic charge distribution. Methyl fluoride was one of the first molecules to be studied at both Doppler limited and sub-Doppler resolution. A series of measurements using both the laser Stark and OODR method [47, 51–53] has allowed the variation of the dipole moment to be measured in some detail. The results are summarized in Table 1.1.

Table 1.1: Dipole Moments (Debye) of CH_3F, $^{13}CH_3F$, and $^{12}CD_3F$ In Various Vibrational States

State	$^{12}CH_3F$	Ref.	$^{13}CH_3F$	Ref.	$^{12}CD_3F$	Ref.
vibr. ground state	1.8585(5)	a	1.8579(6)	b	1.8702(21)	c
ν_3	1.9054(6)	b	1.9039(6)	b	1.8964(15)	c
ν_5					1.8751(21)	e
ν_6	1.859(5)	f			1.8771(7)	d
$2\nu_3$	1.9519(20)	b	1.951(4)		1.9170(5)	e
$\nu_3 + \nu_6$	1.909(5)	f			1.932(7)	f

[a]M. D. Marshall and J. S. Muenter, *J. Mol. Spectrosc.* 83:279, 1980. [b]S. M. Freund, G. Duxbury, M. Romheld, J. T. Tiedje, and T. Oka, *J. Mol. Spectrosc.* 52:38, 1974. [c]G. Duxbury, S. M. Freund, and J. W. C. Johns, *J. Mol. Spectrosc.* 62:99, 1976. [d]G. Duxbury and S. M. Freund, *J. Mol. Spectrosc.* 67:219, 1977. [e]G. L. Caldow and G. Duxbury, *J. Mol. Spectrosc.* 89:93, 1981. [f]G. Duxbury and H. Kato, *Chem. Phys.* 66:161, 1982.

Another well-studied polar molecule is formaldehyde [24, 54–56], see the results in Tables 1.2 and 1.3.

Table 1.2: Dipole Moments (Debye) of Formaldehyde and Thioformaldehyde In Various Vibrational and Electronic States

	State	H_2CO	Ref.	H_2CS	Ref.
$^1A_1$	vibrational ground state	2.3315(5)	a	1.6491(4)	a
	$v_3 = 1$ (CS), $v_2 = 1$ (CO)	2.3470(5)	b	1.6576(12)	c
	$v_4 = 1$	2.3086(5)	a	1.622(3)	e
	$v_6 = 1$	2.3285(5)	a	1.642(5)	e
	$v_3 = 1$ (CO)	2.3250(25)	d		
	$v_5 = 1$	2.2844(47)	d		
	$v_2 = 2$	2.3605(20)	d		
	$v_2 = 4$	2.2723(86)	f		
	$v_2 = 2\ v_4 = 2$	2.3222(47)	f		
	$v_2 = 1\ v_4 = 4$	2.2825(33)	f		
$\tilde{a}^3A_2$		1.29(3)	g		
$\tilde{A}^1A_2$		1.56(7)	g	0.79(4)	c

[a]B. Fabricant, D. Krieger, and J. S. Muenter, *J. Chem. Phys.* 67:1576, 1977. [b]C. Brechignac, J. W. C. Johns, A. R. W. McKellar, and M. Wong, *J. Mol. Spectrosc.* 96:353, 1982. [c]D. J. Bedwell and G. Duxbury, *J. Mol. Spectrosc.* 84:531, 1980. [d]M. Allegrini, J. W. C. Johns, and A. R. W. McKellar, *J. Mol. Spectrosc.* 67:476, 1977. [e]G. Duxbury, H. Kato, and M. L. Le Lerre, *Faraday Discuss. Chem. Soc.* 71:97, 1981. [f]P. H. Vaccaro, J. L. Kinsey, R. W. Field, and H. L. Dai, *J. Chem. Phys.* 78:3659, 1983. [g]R. N. Dixon and C. R. Webster, *J. Mol. Spectrosc.* 70:314, 1978.

Table 1.3: Dipole Moments (Debye) of Deuterated Formaldehyde and Deuterated Thioformaldehyde In Various Vibrational States

State		D_2CO	Ref.	D_2CS	Ref.
$^1A_2$	vibrational ground state	2.3471(5)	d	1.658(3)	e
	$v_3 = 1$ (CS), $v_2 = 1$ (CO)	2.3672(15)	d	1.658(3)	e
	$v_3 = 1$ (CO)	2.319(10)	d		
	$v_6 = 1$	2.347(4)	d		

^{d}M. Allegrini, J. W. C. Johns, and A. R. W. McKellar, *J. Mol. Spectrosc.* 67:476, 1977. ^{e}G. Duxbury, H. Kato, and M. L. Le Lerre, *Faraday Discuss. Chem. Soc.* 71:97, 1981.

In order to observe OODR spectra, very small radiofrequencies and very high values of the applied electrostatic field are required. An OODR spectrum of HNO is shown in Figure 1.12, and the corresponding energy diagram in Figure 1.13.

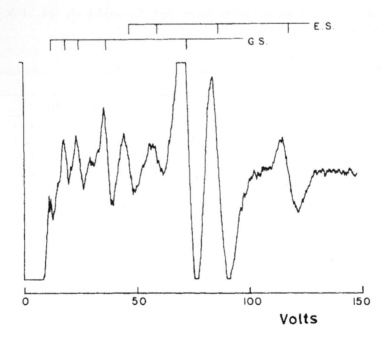

Figure 1.12: M_J resolved OODR spectra of HNO using laser-induced fluorescence, with the difference frequency generated using an amplitude modulated acousto-optic modulator: mixed polarization spectrum of the $^RR_3(3)$ line of the $A^1A''(100) - X^1A'(000)$ band with $\Omega_{AM} = 15\,\text{MHz}$, and a 3 mm spacing between the Stark electrodes. ES, excited state resonances, and GS, ground state resonances. Reprinted from Dixon and Noble, *Chem. Phys.* 50:331, 1980 [57]. Copyright (1980), with permission from Elsevier.

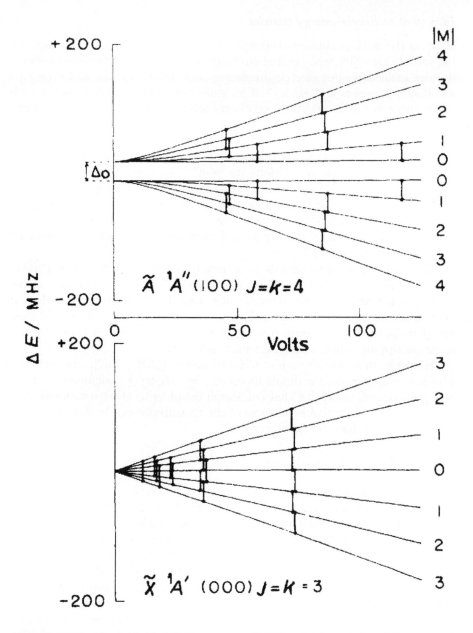

Figure 1.13: M_J resolved OODR spectra of HNO using laser induced fluorescence: energy level diagram showing the Stark splitting and resonance intervals corresponding to the spectrum in Figure 1.12. Reprinted from Dixon and Noble, *Chem. Phys.* 50:331, 1980 [57]. Copyright (1980), with permission from Elsevier.

Effects of collisional energy transfer

Much of the work on collisional energy transfer has stemmed from the pioneering work of Oka [58], who pointed out that many collisional processes between dipolar molecules obey electric dipole selection rules, i.e., the molecule after a collision possesses a "memory" of its state before the collision. In the early work, there was little evidence for velocity selection effects, since the majority of the experiments were carried out in the microwave region where the broadening processes are largely homogeneous [58]. However, in a set of infrared experiments using two-photon pumping and probing, Oka and his colleagues showed that many dipolar collisions are "soft" [59]. This means that, although the molecules change their rotational quantum state following electric dipole selection rules, the velocity component in the direction of the laser field is unchanged. Following this difficult experiment, it was realised that these effects could be seen "routinely" in laser Stark Lamb dip, and in OODR experiments as "four-level resonances."

An example of these signals is shown in Figure 1.14 for thioformaldehyde and illustrates that four-level collisionally transferred resonances may approach at least 50% of the intensity of a three-level resonance. This demonstrates that in molecules in which the dipole moment is directed entirely along the principal near symmetric rotor axis, a or c, the $\Delta M = \pm 1$ angular momentum tipping collisions are extremely selective.

However, in molecules such as CH_3OH and CH_2NH [61,62], where there is also a b component of the dipole moment, the four-level resonances are much less pronounced, indicating that collisional coupling to other rotational levels is now much more probable. In some of the transitions seen in H_2CO, changes of $\Delta M = 4$ or 6 have been observed [21].

1.2.5 Studies of Semistable Molecules

One of the principal advantages of laser spectrometers is their sensitivity for the detection of small quantities of short-lived species. This sensitivity has been exploited in the study of small "semistable" molecules which, in Kroto's definition [63], have lifetimes of the order of seconds under the conditions of the gas phase experiments. For example, the laser Stark spectroscopy method is suitable for studying molecules such as HN0, CH_2NH, and H_2CS, but not most free radicals, since the metal surfaces of the Stark plates catalyse the decomposition of the unstable molecules.

One of the most interesting of the semistable molecules to be studied by laser methods is thioformaldehyde, H_2CS. Table 1.4 gives an overview of theoretical and experimental data. In fact, thioformaldehyde can now join its prototype, formaldehyde, as one of the best characterized tetra-atomic molecules.

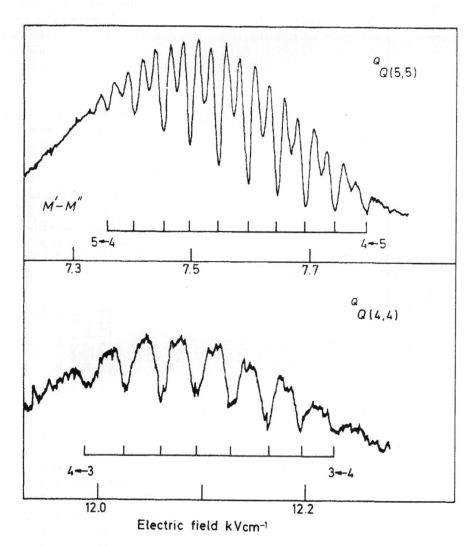

Figure 1.14: Lamb dip spectra of the $^QQ(5,5)$ and the $^QQ(4,4)$ transitions of the ν_3 band of H_2CS using second derivative presentation. The four-level collisionally transferred resonances seen between Lamb dips of the $^QQ(5,5)$ transition are approximately 50% of the intensity of the Lamb dips. The time constant for detection was 300 ms and the sample pressure about 5 mtorr. Reprinted from Bedwell and Duxbury, *J. Mol. Spectrosc.* 84:531, 1980 [60]. Copyright (1980), with permission from Elsevier.

Table 1.4: Comparison of Observed and Calculated Vibrational Fundamentals of Thioformaldehyde with Those in $\tilde{A}^1A_2$ and $\tilde{a}^3A_2$ States of Thioformaldehyde

Vibration	Description	Ref.	$H_2CO(\tilde{X}^1A_1)$	$H_2CS(\tilde{X}^1A_1)$	$H_2CS(\tilde{A}^1A_2)$	$H_2CS(\tilde{a}^3A_2)$
$\nu_1\ A_1$	symmetric CH stretch	a	2782	2971	3034	
		b	2782	3057		2962
		c		2937		
$\nu_2\ A_1$	symmetric HCH bend	a	1500	1439	1310	
		b	1503	1440		1320
		c		1464		1346
$\nu_3\ A_1$	CS (CO) stretch	a	1746	1059	820	
		b	1709	1058		861.6
		c		1053		792
$\nu_4\ B_1$	out-of-plane bend	a	1167	990	371	
		b	1161	1065		356
		c		1029		383
$\nu_5\ B_2$	anti-symmetric CH stretch	a	2843	3025	3081	
		b	2884	3054		3078
		c		3023		
$\nu_6\ B_2$	HCH wag	a	1249	991	799	
		b	1245	989		762.3
		c		968		761

^{a}Experimental -H$_2$CO ref. b H$_2$CS D. J. Bedwell and G. Duxbury, *J. Mol. Spectrosc.* 84:53, 1980. W. B. Olsen and J. W. C. Johns, *J. Mol. Spectrosc.* 39:47, 1971. D. J. Clouthier, C. M. L. Kerr, and D. A. Ramsey, *Chem. Phys.* 56:73, 1981. R. H. Judge and G. W. King, *J. Mol. Spectrosc.* 74:175, 1979.
^{b}Theoretical frequencies from R. Jaquet, W. Kutzelnigg, and V. Staemmler, *Theor. Chem. Acta.* 54:205, 1980. Harmonic frequencies reduced by 5.15% to give correct weighting for H$_2$CO, for justification for scaling see ref. c. ^{c}Theoretical frequencies from J. D. Goddard and D. J. Clouthier, *J. Chem. Phys.* 76:5039, 1982. 10.5% scaling of calculated harmonic frequencies applied.

1.2.6 Spectroscopic Studies of Free Radicals and Molecular Ions

The development of laser-based methods for the study of free radical spectra has paralleled that for stable and semi-stable molecules. The first methods to be developed were of the fixed frequency type, in particular, LMR, the laser analogue of gas-phase electron paramagnetic resonance spectroscopy. Following the development of tunable visible and infrared lasers, many free radicals have been studied via their electronic and vibration-rotation spectra. Two representative examples are BO_2 and CH_2, shown in Figures 1.15 and 1.16.

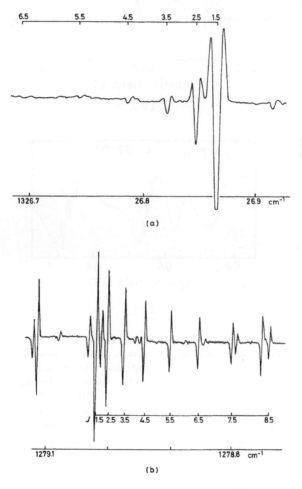

Figure 1.15: (a) Q-branch transitions of the ν_3 band of $^{10}BO_2$ in the $^2\Pi_{3/2}$ state, obtained by Zeeman modulation. (b) Analogous transitions of $^{11}BO_2$ obtained by frequency modulation of the diode. From Kawaguchi et al., *Mol. Phys.* 44:509, 1981 [64], reprinted by permission of the publisher (Taylor & Francis Ltd).

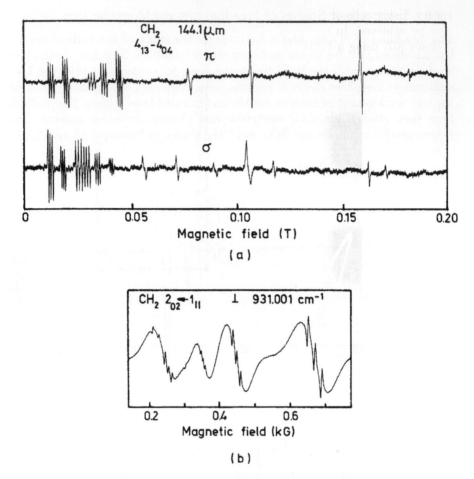

Figure 1.16: LMR spectra of CH_2, using far infrared (FIR) and mid-infrared lasers, showing triplet hyperfine structure. (a) Spectra using the 144.1 μm (69.38765 cm^{-1}) laser line of CD_3OH in π and σ polarizations. The resonances observed are for various Zeeman components of the $4_{13} - 4_{04}$ rotational transition. (b) Spectrum observed for the $^{12}C^{16}O_2$ $P(34)$ laser line at 931.001 cm^{-1} due to the $2_{02} - 1_{11}$ transition. The triplet structure is revealed as a result of saturation spectroscopy (Lamb dips). Reproduced from Sears et al., *J. Chem. Phys.* 77:5348, 1982, and 77:5363, 1982 [65,66], with permission from AIP Publishing.

1.2.7 Ultra-High-Resolution Spectroscopy

Spectra obtained using noble gas and CO_2 waveguide laser-based spectrometers are among some of the highest-resolution spectra ever obtained. In order to achieve such high resolution, very-large-diameter cells of up to 38 cm have been used, and in order to use low pressures, cell lengths of up to 13 m are necessary. Using cells of this type wavefronts approaching those of a plane wave can be achieved. In one of the pioneering experiments by Hall, Bordé, and Uehara [67] using a HeNe laser, the recoil splitting of hyperfine components in CH_4 was observed; see Figure 1.17.

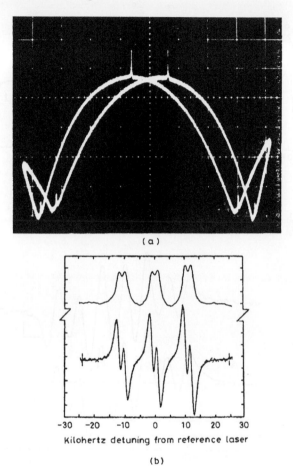

Figure 1.17: (a) Saturation peak at the output of a He-Ne laser seen in the original intracavity "inverted" Lamb dip experiment on CH_4. The laser frequency was swept twice over the gain profile. (b) Ultra-high-resolution spectrum of the three main hyperfine lines of $^{12}CH_4$ showing the recoil doubling (lower curve). Methane pressure 70 μtorr; room temperature; modulation of 800 Hz peak to peak. A least-squares fit (solid line) gives a width of 1.27 kHz half-width at half-maximum (HWHM) and a recoil doublet splitting of 2.150 kHz, the high frequency peak 1% larger than the low frequency ones. The upper curve, integrated from a sample of such data, shows that each hyperfine component is spectrally doubled by the recoil effect. Reprinted figures with permission from Hall and coworkers, *Phys. Rev. Lett.* 22:4, 1969, and 37:1339, 1976 [67,68]. Copyright (1969) and (1976) by the American Physical Society.

Another example of an ultra-high-resolution spectrum is that of SF_6 shown in Figure 1.18, where the half-width at half maximum of 5 kHz corresponds to a resolution of about 1 part in 10^{10}.

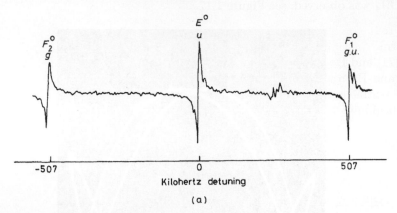

(a)

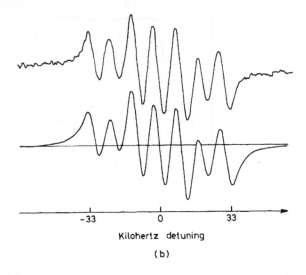

(b)

Figure 1.18: (a) Ultra-high-resolution spectrum of the SF_6 Q_{38} cluster at 28.412582452 THz recorded using a frequency-controlled saturation based on a waveguide CO_2 laser. (b) Magnetic hyperfine structure of the $R_{28}A_2^0$ line of $^{32}SF_6$ at 28.46469125 THz. The upper trace is the observed spectrum, and the lower trace calculated using the solar spin-rotation interaction $W_{SRS} = -hc_aIJ$, with $c_a = -5$ kHz. The resolution is ca. 5 kHz. Reprinted (a) from Ch. J. Bordé and coworkers, *J. Mol. Spectrosc.* 73:344, 1978 [69]. Copyright (1980), with permission from Elsevier. Reprinted figure (b) with permission from J. Bordé and coworkers, *Phys. Rev. Lett.* 45:14, 1980 [70]. Copyright (1980) by the American Physical Society.

1.2.8 Conclusion

In this chapter, we have endeavoured to present a picture of the impact of lasers on high-resolution gas phase molecular spectroscopy by concentrating on certain key areas. This approach, by its very nature, must lead to several topics being either omitted, or treated in a perfunctory fashion. In-depth treatment of much of the subject matter has been given in the books by Hollas [71] and Demtröder [29] and in the excellent review articles of Macpherson and Barrow [72, 73]. In these reviews, an almost complete bibliography of experimental work in gas phase molecular spectroscopy from 1978 up to the beginning of 1982 has been given. Further development in high-resolution spectroscopy since these early days, with focus on dihydrides, is described in Chapters 4 to 6 of the present work. In Chapter 7, examples of measurements from the Herschel Space Observatory are presented.

1.3 Bibliography

[1] H. Sponer and E. Teller. Electronic spectra of polyatomic molecules. *Reviews of Modern Physics*, 13:75–170, Apr 1941.

[2] R. G. W. Norrish and G. Porter. Chemical reactions produced by very high light intensities. *Nature*, 164:658–658, 1949.

[3] G. Porter. Flash photolysis and spectroscopy. A new method for the study of free radical reactions. *Proceedings of the Royal Society of London. Series A. Mathematical and Physical Sciences*, 200(1061):284–300, 1950.

[4] G. Porter. Nobel Lecture: Flash photolysis and some of its applications. *Science*, 160(3834):1299–1307, 1968.

[5] G. Porter and M. R. Topp. Nanosecond flash photolysis. *Proceedings of the Royal Society of London. A. Mathematical and Physical Sciences*, 315(1521):163–184, 1970.

[6] G. Porter and M. R. Topp. Nanosecond flash photolysis and the absorption spectra of excited singlet states. *Nature*, 220:1228–1229, 1968.

[7] T. G. Kim and M. R. Topp. Femtosecond fluorescence up-conversion spectrometer. *Journal of Physical Chemistry A*, 108:10060–10065, 2004.

[8] G. Herzberg and D. A. Ramsay. Absorption spectrum of free NH_2 radicals. *The Journal of Chemical Physics*, 20(2):347–347, 1952.

[9] K. Dressler and D. A. Ramsay. Renner effect in polyatomic molecules. *The Journal of Chemical Physics*, 27(4):971–972, 1957.

[10] J. A. Pople and H. C. Longuet-Higgins. Theory of the Renner effect in the NH_2 radical. *Molecular Physics*, 1(4):372–383, 1958.

[11] K. Dressler and D. A. Ramsay. The electronic absorption spectra of NH_2 and ND_2. *Philosophical Transactions of the Royal Society of London. Series A, Mathematical and Physical Sciences*, 251:553–602, 1959.

[12] R. N. Dixon. The absorption spectrum of the free NCO radical. *Philosophical Transactions of the Royal Society of London. Series A, Mathematical and Physical Sciences*, 252(1007):165–192, 1960.

[13] R. N. Dixon. A $^2\Pi - ^2\Pi$ electronic band system of the free NCO radical. *Canadian Journal of Physics*, 38(1):10–16, 1960.

[14] G. Herzberg. Twelfth Spiers memorial lecture. determination of the structures of simple polyatomic molecules and radicals in electronically excited states. *Discussions of the Faraday Society*, 35:7–29, 1963.

[15] G. Herzberg and J. W. C. Johns. The spectrum and structure of singlet CH_2. *Proceedings of the Royal Society of London. Series A. Mathematical and Physical Sciences*, 295(1441):107–128, 1966.

[16] G. Duxbury. High resolution laser spectroscopy. *Chemical Society Reviews*, 12:453–504, 1983.

[17] W. R. Bennett. Hole burning effects in a He-Ne optical maser. *Physical Review*, 126:580–593, 1962.

[18] W. R. Bennett Jr. Hole burning effects in gas lasers with saturable absorbers. *Comments on Atomic and Molecular Physics*, 2:10–18, 1970.

[19] R. A. McFarlane, W. R. Bennett, and W. E. Lamb. Single mode tuning dip in the power output of an He-Ne optical maser. *Applied Physics Letters*, 2(10):189–190, 1963.

[20] P. H. Lee and M. L. Skolnick. Saturated neon absorption inside a 6238-Å laser. *Applied Physics Letters*, 10(11):303–305, 1967.

[21] J. W. C. Johns, A. R. W. McKellar, T. Oka, and M. Römheld. Collision-induced Lamb dips in laser Stark spectroscopy. *The Journal of Chemical Physics*, 62(4):1488–1496, 1975.

[22] H. R. Schlossberg and A. Javan. Saturation behavior of a Doppler-broadened transition involving levels with closely spaced structure. *Physical Review*, 150:267–284, 1966.

[23] R. L. Shoemaker, S. Stenholm, and R. G. Brewer. Collision-induced optical double resonance. *Physical Review A*, 10:2037–2050, 1974.

[24] J. W. C. Johns and A. R. W. McKellar. Stark spectroscopy with the CO laser: The dipole moment of H_2CO in the $v_2 = 2$ state. *The Journal of Chemical Physics*, 63(4):1682–1685, 1975.

[25] R. G. Brewer. Precision determination of CH_3F dipole moment by nonlinear infrared spectroscopy. *Physical Review Letters*, 25:1639–1641, 1970.

[26] M. S. Feld, A. Sanchez, A. Javan, and B. J. Feldman. Méthods de spectroscopie sans largeur Doppler de niveaux excités de systèmes moléculaires simples. *Colloques International du CNRS*, 217:87–104, 1974.

[27] H. Uehara and K. Hakuta. Laser magnetic resonance spectra of ClO_2 tuned by the avoidance of Zeeman level crossing. *Chemical Physics Letters*, 58(2):287–290, 1978.

[28] J. Sakai and M. Katayama. Observation of level anticrossing in nonlinear absorption. *Applied Physics Letters*, 28(3):119–121, 1976.

[29] W. Demtröder. *Laser Spectroscopy: Basic Concepts and Instrumentation*. Springer Verlag, 1981.

[30] V. S. Letokhov. Nonlinear high resolution laser spectroscopy. *Science*, 190(4212):344–351, 1975.

[31] V. S. Letokhov and V. P. Chebotayev. *Nonlinear laser spectroscopy*,

volume 4 of *Springer Series in Optical Science*. Springer Verlag, Berlin, 1977.

[32] F. Shimizu. Theory of two-photon Lamb dips. *Physical Review A*, 10:950–959, 1974.

[33] T. Oka. Infrared and radiofrequency spectroscopy in the laser cavity. In S. L. R. Balian, S. Haroche, editor, *Frontiers in Laser Spectroscopy, II*, Les Houches Summer School Proceedings XXVII, pages 531–569. North Holland, Amsterdam, 1977.

[34] C. H. Townes and A. L. Schawlow. *Microwave Spectroscopy*. McGraw-Hill, 1955.

[35] K. Sakurai, K. Uehara, M. Takami, and K. Shimoda. Stark effect of vibration-rotation lines of formaldehyde observed by a 3.5 μm laser. *Journal of the Physical Society of Japan*, 23(1):103–109, 1967.

[36] F. Shimizu. Stark spectroscopy of $^{15}NH_3$ ν_2 band by 10-μ lasers. *The Journal of Chemical Physics*, 53:1149–1151, 1970.

[37] F. Shimizu. Stark spectroscopy of NH_3 ν_2 band by 10-μ CO_2 and N_2O lasers. *The Journal of Chemical Physics*, 52:3572–3576, 1970.

[38] T. E. Gough, R. E. Miller, and G. Scoles. Sub-Doppler resolution infrared molecular-beam spectroscopy. Stark effect measurement of the dipole moment of hydrogen fluoride and hydrogen cyanide in excited vibrational states. *Faraday Discussions of the Chemical Society*, 71:77–85, 1981.

[39] J. Orr and T. Oka. Doppler-free optical double resonance spectroscopy using a single-frequency laser and modulation sidebands. *Applied Physics*, 21(4):293–306, 1980.

[40] D. J. Bedwell and G. Duxbury. Single sideband Doppler-free double resonance Stark spectroscopy using an acousto-optic modulator. *Chemical Physics*, 37(3):445–452, 1979.

[41] K. M. Evenson, H. P. Broida, J. S. Wells, R. J. Mahler, and M. Mizushima. Electron paramagnetic resonance absorption in oxygen with the HCN laser. *Physical Review Letters*, 21:1038–1040, 1968.

[42] J. S. Wells and K. M. Evenson. A new LEPR spectrometer. *Review of Scientific Instruments*, 41(2):226–227, 1970.

[43] P. B. Davies, D. K. Russell, B. A. Thrush, and H. E. Radford. Analysis of the laser magnetic resonance spectra of NH_2 ($\tilde{X}^2 B_1$). *Proceedings of the Royal Society of London A: Mathematical, Physical and Engineering Sciences*, 353(1674):299–318, 1977.

[44] A. Carrington. Review lecture: Spectroscopy of molecular ion beams. *Proceedings of the Royal Society of London A: Mathematical, Physical and Engineering Sciences*, 367(1731):433–449, 1979.

[45] J. W. C. Johns and A. R. W. McKellar. Laser Stark spectroscopy of the fundamental bands of HNO (ν_2 and ν_3) and DNO (ν_1 and ν_2). *The Journal of Chemical Physics*, 66(3):1217–1224, 1977.

[46] G. L. Caldow, G. Duxbury, and L. A. Evans. Laser Stark saturation spectroscopy of the ν_2 band of CD_3I. *Journal of Molecular Spectroscopy*, 69(2):239–253, 1978.

[47] S. M. Freund, G. Duxbury, M. Römheld, J. T. Tiedje, and T. Oka. Laser Stark spectroscopy in the 10 μm region: The ν_3 bands of CH_3F. *Journal of Molecular Spectroscopy*, 52(1):38–57, 1974.

[48] A. Carrington. *Microwave spectroscopy of free radicals*. Academic Press, 1974.

[49] M. S. Sorem and A. L. Schawlow. Saturation spectroscopy in molecular iodine by intermodulated fluorescence. *Optics Communications*, 5(3):148–151, 1972.

[50] C. Wieman and T. W. Hänsch. Doppler-free laser polarization spectroscopy. *Physical Review Letters*, 36:1170–1173, May 1976.

[51] G. Duxbury and S. M. Freund. Stark spectroscopy with the CO_2 laser: The ν_6 band of CD_3F. *Journal of Molecular Spectroscopy*, 67(1–3):219–243, 1977.

[52] G. Duxbury and H. Kato. Optical-optical double resonance spectra of CH_3F and CD_3F using isotopic CO_2 lasers. *Chemical Physics*, 66(1):161–167, 1982.

[53] G. Duxbury, S. M. Freund, and J. W. C. Johns. Stark spectroscopy with the CO_2 laser: The ν_3 band of CD_3F. *Journal of Molecular Spectroscopy*, 62(1):99–108, 1976.

[54] M. Allegrini, J. W. C. Johns, and A. R. W. McKellar. A study of the Coriolis-coupled ν_4, ν_6, and ν_3 fundamental bands and the $\nu_5 \leftarrow \nu_6$ difference band of H_2CO; Measurement of the dipole moment for $v_5 = 1$. *Journal of Molecular Spectroscopy*, 67(1):476–495, 1977.

[55] D. Coffey Jr., C. Yamada, and E. Hirota. Laser Stark spectroscopy of formaldehyde-d_2: The Coriolis coupled ν_4 and ν_6 fundamentals. *Journal of Molecular Spectroscopy*, 64(1):98–108, 1977.

[56] P. H. Vaccaro, J. L. Kinsey, R. W. Field, and H.-L. Dai. Electric dipole moments of excited vibrational levels in the $\tilde{X}^1 A_1$ state of formaldehyde by stimulated emission spectroscopy. *The Journal of Chemical Physics*, 78(6):3659–3664, 1983.

[57] R. N. Dixon and M. Noble. The dipole moment of HNO in its $\tilde{A}^1 A''$ excited state determined using optical-optical double resonance Stark spectroscopy. *Chemical Physics*, 50(3):331–339, 1980.

[58] T. Oka. Collision-induced transitions between rotational levels. *Advances in Atomic and Molecular Physics*, 9:127–206, 1973.

[59] S. M. Freund, J. W. C. Johns, A. R. W. McKellar, and T. Oka. Infrared-infrared double resonance experiments using a two-photon technique. *The Journal of Chemical Physics*, 59(7):3445–3453, 1973.

[60] D. J. Bedwell and G. Duxbury. Laser Stark spectroscopy of thioformaldehyde in the 10μm region: The ν_3, ν_4, and the ν_6 fundamentals. *Journal of Molecular Spectroscopy*, 84(2):531–558, 1980.

[61] D. J. Bedwell, G. Duxbury, H. Herman, and C. A. Orengo. Laser Stark and optical-optical double resonance studies of some molecules used in optically pumped submillimetre lasers. *Infrared Physics*, 18(5):453–460, 1978.

[62] G. Duxbury, H. Kato, and M. L. Le Lerre. Laser Stark and interferometric studies of thioformaldehyde and methyleneimine. *Faraday Discussions of the Chemical Society*, 71:97–110, 1981.

[63] H. W. Kroto. Tilden Lecture. Semistable molecules in the laboratory and in space. *Chemical Society Reviews*, 11:435–491, 1982.

[64] K. Kawaguchi, E. Hirota, and C. Yamada. Diode laser spectroscopy of the BO_2 radical vibronic interaction between the $\tilde{A}^2\Pi_u$ and $\tilde{X}^2\Pi_g$ states. *Molecular Physics*, 44(2):509–528, 1981.

[65] T. J. Sears, P. R. Bunker, A. R. W. McKellar, K. M. Evenson, D. A. Jennings, and J. M. Brown. The rotational spectrum and hyperfine structure of the methylene radical CH_2 studied by far-infrared laser magnetic resonance spectroscopy. *The Journal of Chemical Physics*, 77(11):5348–5362, 1982.

[66] T. J. Sears, P. R. Bunker, and A. R. W. McKellar. The laser magnetic resonance spectrum of the ν_2 band of the methylene radical CH_2. *The Journal of Chemical Physics*, 77(11):5363–5369, 1982.

[67] J. L. Hall, Ch. J. Bordé, and K. Uehara. Direct optical resolution of the recoil effect using saturated absorption spectroscopy. *Physical Review Letters*, 37:1339–1342, November 1976.

[68] R. L. Barger and J. L. Hall. Pressure shift and broadening of methane line at 3.39 μ studied by laser-saturated molecular absorption. *Physical Review Letters*, 22:4–8, January 1969.

[69] Ch. J. Bordé, M. Ouhayoun, and J. Bordé. Observation of magnetic hyperfine structure in the infrared saturation spectrum of $^{32}SF_6$. *Journal of Molecular Spectroscopy*, 73(2):344–346, 1978.

[70] J. Bordé, Ch. J. Bordé, C. Salomon, A. Van Lerberghe, M. Ouhayoun, and C. D. Cantrell. Breakdown of the point-group symmetry of vibration-rotation states and optical observation of ground-state octahedral splittings of $^{32}SF_6$ using saturation spectroscopy. *Physical Review Letters*, 45:14–17, July 1980.

[71] J. M. Hollas and W. Jeremy Jones. *High resolution spectroscopy*. Wiley Online Library, 1983.

[72] M. T. Macpherson and R. F. Barrow. Gas-phase molecular spectroscopy. *Annual Reports on the Progress of Chemistry, Section C: Physical Chemistry*, 76:51–98, 1979.

[73] M. T. Macpherson and R. F. Barrow. Gas-phase molecular spectroscopy. *Annual Reports on the Progress of Chemistry, Section C: Physical Chemistry*, 78:221–376, 1981.

CHAPTER 2

Linear and Bent Molecule Vibration-Rotation Hamiltonians for Open-Shell Molecules

CONTENTS

In this chapter, we discuss, following Duxbury [1], effective Hamiltonians of the type described by Van Vleck [2], where all contributions that are off-diagonal in vibrational or electronic quantum numbers in a particular basis set have been absorbed into the parameters of the effective Hamiltonian. These parameters can then be compared with those calculated from the full molecular Hamiltonian.

2.1   LINEAR TRIATOMIC MOLECULES

The treatment of the rotational energy levels of linear triatomic molecules proceeds using the methods discussed by Hougen [3] for diatomic molecules. The differences arise primarily because triatomic molecules possess a bending vibration that destroys the linear symmetry.

In a linear triatomic molecule, there are four vibrational degrees of freedom, two of which are associated with the degenerate bending vibration, and two rotational degrees of freedom. Linear molecules possess a vanishing moment of inertia about the linear axis, and a nonvanishing moment of inertia perpendicular to the linear axis, which is the same for all such axes through the centre of mass. Consequently, rotational levels are characterised by a single rotational constant, the B value in the rigid-rotor approximation.

The linear axis provides a unique direction in the molecule along which projections of the various angular momenta can be quantised. Thus, in linear triatomic molecules, we must consider the projection l of the vibrational angular momentum along this axis. If the molecule is in a degenerate electronic state, we must also consider the projection Λ of the total electronic angular momentum L, and if it is a multiplet state, the projection Σ of the total spin angular momentum S.

2.1.1   Basis Functions and Operators

Basis set functions for a linear triatomic molecule can be written in the form $|L\Lambda S\Sigma\rangle|v_1 v_2^l v_3\rangle|PJM\rangle$. In the electronic part, L is often such a poor quantum number that it is omitted from the basis set. The vibrational part is characterised by four quantum numbers, of which v_1 and v_3 refer to the stretching vibrations, and v_2 and l to the degenerate bending vibration. The rotational part of the basis set function is characterised by $P \equiv \Lambda + \Sigma + l$, J, and M, where J and M represent the total angular momentum of the molecule and its projection along the space-fixed axis. Following Hougen [3], we write the total Hamiltonian

$$H = H_{ev} + H_r, \tag{2.1}$$

where H_{ev} is the vibronic part of the Hamiltonian excluding the rotational variables. H_{ev} is sometimes called the Hamiltonian for the nonrotating molecule.

2.1.2   The Nonrotating Molecule

If the Renner effect is neglected, H_{ev} can be written as

$$H_{ev} = H_{ev}^0 + H_{so}, \tag{2.2}$$

where H_{ev}^0 is the ordinary vibrational and electronic Hamiltonian in the absence of any vibronic or spin-orbit interaction, and H_{so} is the spin-orbit part.

H_{ev}^0 is diagonal in our basis set and has nonvanishing matrix elements given to a first approximation by

$$\langle \Lambda S \Sigma | \langle v_1 v_2^l v_3 | \langle PJM | H_{ev}^0 | PJM \rangle | v_1 v_2^l v_3 \rangle | \Lambda S \Sigma \rangle =$$

$$E_e + \left(v_1 + \frac{1}{2} \right) \omega_1 + \left(v_2 + \frac{1}{2} \right) \omega_2 + \left(v_3 + \frac{1}{2} \right) \omega_3, \qquad (2.3)$$

where E_e is a constant electronic energy for the electronic state under consideration, and the ω_i are its vibration frequencies. The second operator, H_{so}, is the spin-orbit interaction operator, which for linear molecules is written as $\boldsymbol{AL} \cdot \boldsymbol{S}$. It has diagonal nonvanishing matrix elements

$$\langle \Lambda S \Sigma | \langle v_1 v_2^l v_3 | \langle PJM | \boldsymbol{AL} \cdot \boldsymbol{S} | PJM \rangle | v_1 v_2^l v_3 \rangle | \Lambda S \Sigma \rangle = \zeta \Lambda \Sigma. \qquad (2.4)$$

This operator also has matrix elements off-diagonal in Λ, occurring with $\Delta \Lambda = \pm 1$. These need to be considered when perturbations by other electronic states occur.

2.1.3 The Rotating Molecule

The quantum mechanical operator of a rigid rotating molecule has the form

$$H_r = B \left(R_x^2 + R_y^2 \right), \qquad (2.5)$$

where R represents the angular momentum that arises exclusively from the rotational motion of the nuclei. R is not normally a very good quantum number, and, therefore, it is more convenient to rewrite Equation (2.5) in terms of the other operators.

Since $\boldsymbol{J} = \boldsymbol{L} + \boldsymbol{S} + \boldsymbol{\Pi} + \boldsymbol{R}$, where $\boldsymbol{\Pi}$ represents the total vibrational angular momentum operator, we can rewrite the rotational Hamiltonian as

$$H_r = B \left[\left(J_x - L_x - S_x - \Pi_x \right)^2 + \left(J_y - L_y - S_y - \Pi_y \right)^2 \right]. \qquad (2.6)$$

For the purposes of calculation, it is useful to write H_r as follows:

$$\begin{aligned} H_r = {} & B \left[J^2 - J_z^2 + L^2 - L_z^2 + S^2 - S_z^2 + \Pi^2 - \Pi_z^2 \right] \qquad (2.7) \\ & - B \left[\left(J^+ L_- + J^- L_+ \right) + \left(J^+ S_- + J^- S_+ \right) + \left(J^+ \Pi_- + J^- \Pi_+ \right) \right] \\ & + B \left[\left(L_+ S_- + L_- S_+ \right) + \left(L_+ \Pi_- + L_- \Pi_+ \right) + \left(S_+ \Pi_- + S_- \Pi_+ \right) \right], \end{aligned}$$

where $J^\pm = J_x \pm i J_y$, $L_\pm = L_x \pm i L_y$, $S_\pm = S_x \pm i S_y$, and $\Pi_\pm = \Pi_x \pm i \Pi_y$ are the usual shift operators. In this chapter, the angular momentum is treated as dimensionless.

Equation (2.7) may be written as $H_r = H^0 + H'$. In this form, the operators in the first row form H^0, and have only diagonal elements between the basis functions, whereas the remainder, H', are off-diagonal with respect to the basis functions. Further, Hougen [4] has shown that the contributions of the last three diagonal operators are equal for all J values, and, therefore, we can write these as $\langle L_\perp^2 \rangle$, $\langle \Pi_\perp^2 \rangle$, $\langle S_\perp^2 \rangle$, respectively. In practice, they will contribute to the vibronic origins. The matrix elements of the remaining operators in H_r have quite a simple form, apart from those of the vibrational angular momentum. The operators L and S behave like ordinary angular momentum operators, and their matrix elements have the following form:

$$
\begin{aligned}
\langle S\Sigma|S^2|S\Sigma\rangle &= S(S+1) \\
\langle S\Sigma|S_z|S\Sigma\rangle &= \Sigma \\
\langle S\Sigma \pm 1|S_\pm|S\Sigma\rangle &= [(S \mp \Sigma)(S \pm \Sigma + 1)]^{\frac{1}{2}} .
\end{aligned}
\tag{2.8}
$$

The expressions for L are identical with those for S, with S replaced by L and Σ replaced by Λ. The operator J, which describes the rotation of the molecule-fixed coordinate system with respect to the laboratory system, obeys anomalous commutation rules (Van Vleck [2]), hence, the roles of J^+ and J^- are interchanged. We indicate this by writing $\pm$ as superscripts. The matrix elements involving the J operator have the form

$$
\begin{aligned}
\langle JP|S^2|JP\rangle &= S(S+1) \\
\langle JP|S_z|JP\rangle &= P \\
\langle JP \pm 1|J^\mp|JP\rangle &= [(J \mp P)(J \pm P + 1)]^{\frac{1}{2}} .
\end{aligned}
\tag{2.9}
$$

The operator Π has the following form for a symmetrical linear triatomic molecule [4]

$$
\begin{aligned}
\Pi_x &= Q_y p_a - Q_a p_y \\
\Pi_y &= Q_a p_x - Q_x p_a \\
\Pi_z &= Q_x p_y - Q_y p_x,
\end{aligned}
\tag{2.10}
$$

where Q_a represents the normal coordinate for the Σ_u^+ asymmetric stretching vibration, and Q_x and Q_y the two perpendicular components of the Π_u bending vibration of the XYX molecule. p_a, p_x and p_y are the momenta conjugate to these coordinates. For XYZ molecules, Q_a and p_a must be replaced by a linear combination of Q_1 and Q_3, and of p_1 and p_3, respectively (see Johns [5]). Since $\langle \Pi_\perp^2 \rangle$ is an additive constant, we are only interested in the matrix elements of Π_x and Π_y. If we let

$$
\begin{aligned}
\Pi_\pm &= \Pi_x \pm i\Pi_y \\
&= \pm iQ_a(p_x \pm ip_y) \mp ip_a(Q_x \pm iQ_y)
\end{aligned}
\tag{2.11}
$$

and set

$$
\frac{\omega_a - \omega_2}{2(\omega_2\omega_a)^{\frac{1}{2}}} = \omega_{2a}^+
\tag{2.12}
$$

and

$$
\frac{\omega_a + \omega_2}{2(\omega_2\omega_a)^{\frac{1}{2}}} = \omega_{2a}^-,
\tag{2.13}
$$

then the matrix elements of Π_- are given in Table 2.1, and evaluated as described by Moffitt and Liehr [6] and Johns [5]. Complete expressions for the more general case of an XYZ molecule are given by Johns [5].

Table 2.1: Matrix elements of the Vibrational Angular Momentum Operator Π_- in the Basis $|v_2^l v_a\rangle$

$\langle v_2'^{-1} v_a' | \Pi_- | v_2^0 v_a \rangle$	$\langle v_2'^{1} v_a' | \Pi_- | v_2^2 v_a \rangle$	v_2'	v_a'
$\sqrt{(v_a+1)(v_2+2)}\,\omega_{2a}^-$	$\sqrt{(v_a+1)v_2}\,\omega_{2a}^-$	v_2+1	v_a+1
$-\sqrt{(v_a+1)v_2}\,\omega_{2a}^+$	$-\sqrt{(v_a+1)(v_2+2)}\,\omega_{2a}^+$	v_2-1	v_a+1
$\sqrt{v_a(v_2+1)}\,\omega_{2a}^+$	$\sqrt{v_a v_2}\,\omega_{2a}^+$	v_2+1	v_a-1
$-\sqrt{v_a v_2}\,\omega_{2a}^-$	$-\sqrt{v_a(v_2+2)}\,\omega_{2a}^-$	v_2-1	v_a-1

The Π_+ matrix elements are given by $\langle v_2'^{-l+1} v_a' | \Pi_+ | v_2^{-l} v_a \rangle = -\langle v_2'^{l-1} v_a' | \Pi_- | v_2^l v_a \rangle$.

Let us consider Equation (2.7) in more detail, starting with a singlet state when $S = 0$. The vibronic angular momentum projection is then

$$\Lambda + l = K, \qquad (2.14)$$

and so the rotational energy in the absence of any interactions is

$$E_r = B\left[J(J+1) - K^2\right]. \qquad (2.15)$$

The remaining terms we need to consider are

$$H' = H_1' + H_3' = -B(J^+ L_- + J^- L_+) - B(J^+ \Pi_- + J^- \Pi_+). \qquad (2.16)$$

H_1' gives rise to l-type doubling, so that H' is the perturbation Hamiltonian for K-type doubling in linear molecules. The other S-independent term, $H_5' = B(L^+ \Pi_- + L^- \Pi_+)$, which has matrix elements $\Delta\Lambda = \pm 1, \Delta l = \mp 1$, can often be neglected.

In multiplet states, we need to consider the effect of the remaining perturbation operators

$$\begin{aligned} H_2' &= -B(J^+ S_- + J^- S_+) \\ H_4' &= B(L_+ S_- + L_- S_+) \\ H_6' &= B(S_+ \Pi_- + S_- \Pi_+). \end{aligned} \qquad (2.17)$$

H_2' is the operator for spin uncoupling due to Coriolis forces and will play an important role in the later discussion of the rotational Hamiltonian in Renner molecules. H_4' is usually small compared with the off-diagonal elements of $\boldsymbol{AL} \cdot \boldsymbol{S}$ and can usually be included as a small correction term to the effective A value. H_6' represents the spin uncoupling due to the vibrational angular momentum, and can be neglected in our examples. In the absence of vibronic interaction, the various formulae or matrices used for diatomic problems can be used for linear triatomic molecules (Hougen [3,4]). The effects of centrifugal distortion can also be added, as described by Hougen [4].

2.2   BENT TRIATOMIC MOLECULES

In bent molecules, we have three vibrational degrees of freedom and three rotational degrees of freedom. Since nonlinear molecules have three nonvanishing moments of inertia, the energy levels of a linear molecule are complicated functions of the three rotational constants, A, B, and C. However, since many of the molecules studied are close to the prolate top limit, $A > B \approx C$, we can often characterize the energy levels approximately by J and K, where $K \equiv K_a$ represents the projection of the total angular momentum along the near prolate top axis. Thus, the near prolate top axis, a, plays the same role for quantization as does the molecular axis in a linear molecule.

2.2.1   Basis Functions and Operators: The Nonrotating Molecule

Basis set functions for a bent triatomic molecule can be written in the form $|LAS\Sigma\rangle|v_1v_2v_3\rangle|PJM\rangle$, where, as in the linear example, L is often omitted from the basis set. The vibrational quantum numbers v_1 and v_3 refer to the stretching vibrations, and v_2 to the bending vibration. The rotational part of the basis function is characterized by $P = K + \Sigma$, so that for singlet states, $|KJM\rangle$ are the symmetric rotor functions implied by the introduction above. The eigenvalues of H_{ev} are now given by

$$\langle AS\Sigma|\langle v_1v_2v_3|\langle PJM|H_{ev}|PJM\rangle|v_1v_2v_3\rangle|AS\Sigma\rangle =$$
$$E_e + \left(v_1 + \frac{1}{2}\right)\omega_1 + \left(v_2 + \frac{1}{2}\right)\omega_2 + \left(v_3 + \frac{1}{2}\right)\omega_3. \qquad (2.18)$$

2.2.2   The Rotating Molecule: Singlet States

The rotational Hamiltonian for an asymmetric rotor in a singlet state, including the effects of centrifugal distortion, has been given in various forms, the most useful of which is due to Watson [7,8]. This is to the fourth power in the rotational angular momentum:

$$\begin{aligned}
H_r = {} & AJ_a^2 + \frac{1}{2}(B+C)\left(J^2 - J_a^2\right) - D_K J_a^4 - D_{JK}J^2\left(J^2 - J_a^2\right) - D_J J^4 \\
& + \frac{1}{2}(B-C)\left(J_b^2 - J_c^2\right) - 2\delta_J J^2\left(J_b^2 - J_c^2\right) \\
& - \delta_K\left[J_a^2\left(J_b^2 - J_c^2\right) + \left(J_b^2 - J_c^2\right)J_a^2\right], \qquad (2.19)
\end{aligned}$$

which has the matrix elements

$$
\begin{aligned}
\langle JK|H_r|JK\rangle &= AK^2 + \frac{1}{2}(B+C)\left[J(J+1)-K^2\right] \\
&\quad - D_K K^4 - D_{JK}J(J+1)K^2 - D_J J^2 (J+1)^2 \quad (2.20) \\
\langle JK|H_r|JK\pm 2\rangle &= \left[\frac{1}{4}(B-C) - \delta_J J(J+1) - \frac{1}{2}\delta_K\left[K^2 + (K\pm 2)^2\right]\right] \\
&\quad \times \sqrt{J(J+1)-K(K\pm 1)} \\
&\quad \times \sqrt{J(J+1)-K(K\pm 1)(K\pm 2)}. \quad (2.21)
\end{aligned}
$$

The matrix elements of the rigid rotor part can be easily derived from those given in Equation (2.9), setting $P = K$.

2.2.3 The Rotating Molecule: Multiplet Electronic States

The two main methods of calculating the matrix elements of asymmetric rotors in multiplet electronic states employ the coupled and the uncoupled representations. The coupled basis was used by Van Vleck [2], Raynes [9] and Bowater et al. [10], and the uncoupled basis by Creutzberg and Hougen [11] and di Lauro [12]. Within these two approaches, the choice can be made of expressing the multiplet part of the Hamiltonian in terms of spherical harmonics, or in a Cartesian basis. Bowater et al. [10] have expressed the entire spin and rotational Hamiltonian in terms of irreducible tensors, which is related to the spherical harmonic approach.

The coupled basis is probably more useful for asymmetric rotors close to the case (b) limit, such as SO_2. However, the uncoupled basis is more useful for molecules, such as AsH_2 and PH_2, which are far from the case (b) limit in vibrationally excited levels of the $^2A_1$ state, and which demonstrate the analogies between the treatment of multiplet states in linear, bent, and quasilinear molecules.

Coupled basis: doublet states

In the coupled Hund's case (b) basis $|NJSK\rangle$, the nonrigid rotor effective Hamiltonian can be written as

$$
H = H_r + H_{sr}, \quad (2.22)
$$

where H_r is given by Equation (2.19), with J replaced by N, the total angular momentum excluding electron spin, and H_{sr} is

$$
H_{sr} = \frac{1}{2}\sum_{\alpha,\beta}\epsilon_{\alpha\beta}(N_\alpha S_\beta + S_\beta N_\alpha) + \frac{1}{2}\sum_{\alpha,\beta,\gamma,\delta}\eta_{\alpha\beta\gamma\delta}(N_\alpha N_\beta N_\gamma S_\delta + S_\delta N_\gamma N_\beta N_\alpha)
$$

$$(2.23)$$

after Lin [13], Dixon and Duxbury [14], Duxbury [15] and Woodman [16]. If we neglect the skew-symmetric terms discussed by Woodman, which have so far not been evaluated in electronic spectra, and if we restrict the η to the one symmetric rotor correction used in the molecules studied so far, the Hamiltonian becomes

$$H_{sr} = \epsilon_{aa} N_a S_a + \epsilon_{bb} N_b S_b + \epsilon_{cc} N_c S_c + \eta_{aaaa} N_a^3 S_a \tag{2.24}$$

plus small correction terms due to the noncommutation of N and S, which can be absorbed into the effective constants in the remainder of the Hamiltonian. The matrix elements of H_{sr} are

$$\langle NK|H_{sr}|NK\rangle = \tag{2.25}$$
$$\frac{J(J+1) - N(N+1) - S(S+1)}{2N(N+1)} \left(\epsilon_{aa} + \eta_{aaaa} K^2 \right) K^2$$
$$+\frac{1}{2} \left(\epsilon_{bb} + \epsilon_{cc} \right) \left[N(N+1) - K^2 \right]$$

$$\langle NK|H_{sr}|NK \pm 2\rangle = \tag{2.26}$$
$$\frac{1}{4} \left(\epsilon_{bb} - \epsilon_{cc} \right) \frac{J(J+1) - N(N+1) - S(S+1)}{2N(N+1}$$
$$\times \sqrt{N(N+1) - K(K \pm 1)} \sqrt{N(N+1) - (K \pm 1)(K \pm 2)}$$

$$\langle NK|H_{sr}|N-1, K\rangle = \tag{2.27}$$
$$\left[\epsilon_{aa} + \eta K^2 - \frac{1}{2} \left(\epsilon_{bb} + \epsilon_{cc} \right) \right] \frac{K}{2N} \sqrt{N^2 - K^2}$$

$$\langle NK|H_{sr}|N-1, K \pm 2\rangle = \tag{2.28}$$
$$\pm \frac{1}{8} \left(\epsilon_{bb} - \epsilon_{cc} \right) \sqrt{(N \mp K)(N \pm K + 1)} \sqrt{(N \mp K - 1)(N \mp K - 2)}.$$

They must be added to those given in Equations (2.20) and (2.21), with J replaced by N, to form the complete spin-molecular rotation matrix.

Coupled basis: triplet states

In triplet states, in addition to spin-rotation interaction, we can have spin-spin interaction, so that

$$H = H_r + H_{sr} + H_{ss}, \tag{2.29}$$

where H_r and H_{sr} have been given in Equations (2.19), (2.23), and (2.24). In a molecule with orthorhombic symmetry, H_{ss} is given by

$$H_{ss} = \alpha \left(2S_z^2 - S_x^2 - S_y^2 \right) + \beta \left(S_x^2 - S_y^2 \right). \tag{2.30}$$

The matrix elements of (2.30) have been given by Van Vleck [2] and Raynes [9]. They are

$$\langle NK|H_{ss}|NK\rangle = \alpha\rho(N)\left[3K^2 - N(N+1)\right] \tag{2.31}$$

$$\langle NK|H_{ss}|NK \pm 2\rangle = \frac{1}{2}\beta\rho(N)\sqrt{N(N+1) - K(K \pm 1)} \tag{2.32}$$
$$\times \sqrt{N(N+1) - (K \pm 1)(K \pm 2)}$$

$$\langle NK|H_{ss}|N-1,K\rangle = \frac{3\alpha}{2}\chi(N)K\sqrt{N^2 - K^2} \tag{2.33}$$

$$\langle NK|H_{ss}|N-1,K \pm 2\rangle = \pm\beta\chi(N)\sqrt{(N \mp K)(N \pm K + 1)} \tag{2.34}$$
$$\times \sqrt{(N \mp K - 1)(N \mp K - 2)},$$

with the auxiliary functions defined as

$$\rho(N) = \frac{3\theta(N)[C(N) + 1] + 2S(S+1)}{(2N-1)(2N+3)}$$

$$\theta(N) = -\frac{C(N)}{2N(N+1)}$$

$$C(N) = J(J+1) - N(N+1) - S(S+1)$$

$$\chi(N) = \frac{[C(N) + N + 1]\phi(N)}{(N-1)(N+1)}$$

$$\phi(N) = -\frac{1}{N}\sqrt{\frac{P(N)Q(N-1)}{(2N-1)(2N+1)}}$$

$$P(N) = (N - J + S)(N + J + S + 1)$$

$$Q(N) = (S + J + N)(N + J - S + 1).$$

Extensions to molecules with lower symmetry have been given by Raynes [9] and Woodman [16].

Uncoupled basis: doublet states

In the uncoupled basis, $|JSPM\Sigma\rangle = |PJM\rangle|S\Sigma\rangle$, the Hamiltonian for rigid rotation and spin-rotation interaction in a doublet state can be written as

$$H = A(J_z - S_z)^2 + B(J_x - S_x)^2 + C(J_y - S_y)^2 + \epsilon_{aa}(J_z - S_z)S_z$$
$$+\epsilon_{bb}(J_x - S_x)S_x + \epsilon_{cc}(J_y - S_y)S_y. \tag{2.35}$$

Following Creutzberg and Hougen [11] and di Lauro [12], this may be rewritten as

$$H = H_1 + H_2 + H_3 + H_4, \tag{2.36}$$

where

$$H_1 = \bar{B}(J^2 - J_z^2) + A(J_z - S_z)^2 + \epsilon_{aa}(J_z - S_z)S_z + \bar{B}(S^2 - S_z^2)$$
$$-\epsilon_{\bar{b}\bar{b}}(S^2 - S_z^2) \tag{2.37}$$

$$H_2 = \left(-\bar{B} + \frac{1}{2}\epsilon_{\bar{b}\bar{b}}\right)(J^+S_- + J^-S_+) \tag{2.38}$$

$$H_3 = \frac{1}{4}(B - C)(J^+J^+ + J^-J^-) \tag{2.39}$$

$$H_4 = \left[-\frac{B - C}{2} + \frac{1}{4}(\epsilon_{bb} - \epsilon_{cc})\right](J^+S_+ + J^-S_-), \tag{2.40}$$

with $\bar{B} = \frac{1}{2}(B + C)$ and $\epsilon_{\bar{b}\bar{b}} = \frac{1}{2}(\epsilon_{bb} + \epsilon_{cc})$. H_1 represents the diagonal matrix elements, H_2 the spin-uncoupling in a symmetric rotor, H_3 the asymmetry splitting, and H_4 represents spin-uncoupling in the asymmetric rotor limit. The matrix elements of these operators are

$$\langle JK\Sigma P|H_1|JK\Sigma P\rangle = \tag{2.41}$$
$$[\bar{B}J(J + 1) - P^2] + AK^2 \pm \epsilon_{aa}K\Sigma + [\bar{B} - \epsilon_{\bar{b}\bar{b}}][S(S + 1) - \Sigma^2]$$

$$\langle JK, \pm\Sigma \pm 1, P \pm 1|H_2|JK, \pm\Sigma, P\rangle = \tag{2.42}$$
$$\left(-B + \frac{1}{2}\epsilon_{\bar{b}\bar{b}}\right)\sqrt{(S + \Sigma)(S - \Sigma + 1)}\sqrt{(J \pm P)(J \mp P + 1)}$$

$$\langle J, K \mp 2, \Sigma, P \mp 2|H_3|JK\Sigma P\rangle = \tag{2.43}$$
$$\frac{1}{4}\sqrt{(J \pm P)(J \mp P + 1)}\sqrt{(J \pm P - 1)(J \mp P + 2)}$$

$$\langle J, K \mp 2, \mp\Sigma \pm 1, P \mp 1|H_4|J, K, \mp\Sigma, P\rangle = \tag{2.44}$$
$$\left[-\frac{1}{2}(B - C) + \frac{1}{4}(\epsilon_{bb} - \epsilon_{cc})\right]\sqrt{(J \pm P - 1)(J \mp P + 2)}.$$

In many of the dihydrides studied so far, the excited states are near prolate symmetric tops, and so most of the energy levels can be described quite well by the matrix elements of H_1 and H_2, together with symmetric top centrifugal distortion corrections to the rotation and the spin-rotation coupling. If $\epsilon_{\bar{b}\bar{b}}$ can be neglected, as is frequently the case, the expression for the rigid rotor energy levels can be rewritten as

$$F(J = K - \frac{1}{2}) = AK^2 + \bar{B}K - \frac{\epsilon}{2} \tag{2.45}$$

$$F(J > K - \frac{1}{2}) = AK^2 + \bar{B}\left[\left(J + \frac{1}{2}\right)^2 - K^2\right]$$
$$\pm \left[4\bar{B}^2\left(J + \frac{1}{2}\right)^2 + \epsilon(\epsilon - 4\bar{B}K)\right]^{\frac{1}{2}}, \tag{2.46}$$

where $\epsilon = \epsilon_{aa}K$. This is analogous to the Hill and Van Vleck [17] expressions

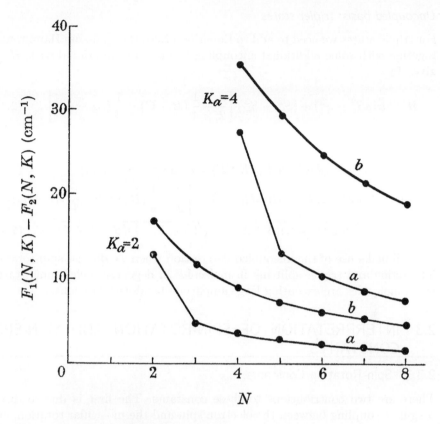

Figure 2.1: Doublet splittings calculated for a symmetric top for $\epsilon_{aa} = 10\,\mathrm{cm}^{-1}$ and $\bar{B} = 4\,\mathrm{cm}^{-1}$, a, including the interaction between the levels $F_1(N, K_a, J)$ and $F_2(N+1, K_a, J)$, and b, neglecting this interaction, i.e., Hund's case (b) coupling. Reproduced by permission from R. N. Dixon et al., Proc. Roy. Soc. A305:271, 1968 [19].

for spin-uncoupling in a diatomic molecule. Thus, for each value of K, we have an effective origin given by AK^2, with a splitting in the absence of overall b and c axis rotation given by $\epsilon_{aa}K \equiv A_{diatomic}$.

The spin-uncoupling can be characterized by the parameter $Y = \epsilon_{aa}K/2\bar{B}$, which plays the same role as $Y = A/2B$ for a diatomic molecule. The effect of large ϵ_{aa} values is a rapid spin-uncoupling from case (a) to case (b) when ϵ_{aa} is positive, as shown in Figure 2.1. In addition, a departure from Hund's case (b) means that many transitions can be seen for which $\Delta N \neq \Delta J$. A detailed analysis was given by di Lauro [18] (see also Reference [19]).

Uncoupled basis: triplet states

For triplet states we need to add to Equation (2.36) the spin-spin Hamiltonian together with some additional uncoupling terms. This additional part, H', is given by

$$H' = \alpha(3S_z^2 - S^2) + (S_+^2 + S_-^2)\left[\frac{\beta}{2} + \frac{1}{4}(B - C) - \frac{1}{4}(\epsilon_{bb} - \epsilon_{cc})\right], \quad (2.47)$$

which has the matrix elements

$$\langle JK\Sigma P|H'|JK\Sigma P\rangle = \alpha\left[3\Sigma^2 - S(S+1)\right] \quad (2.48)$$

$$\langle JK \pm 2, \mp\Sigma \pm 2, P \mp 1|H'|JK\Sigma P\rangle = \left[\frac{\beta}{2} + \frac{1}{4}(B - C) - \frac{1}{4}(\epsilon_{bb} - \epsilon_{cc})\right]$$
$$\times \sqrt{(S \mp \Sigma)(S \pm \Sigma + 1)}\sqrt{(S \mp \Sigma - 1)(S \pm \Sigma + 2)}. \quad (2.49)$$

We will make use of the uncoupled description when we discuss spin-rotation interaction and triplet splitting in molecules in degenerate electronic states, when molecules are executing large amplitude bending vibrations.

2.3 INTERPRETATION OF SPIN-ROTATION AND SPIN-SPIN CONSTANTS

2.3.1 Spin-Rotation Constants, ϵ_{aa}

There are two contributions to these constants. The first is due to direct magnetic coupling between the electron spin and the molecular rotation, and the second to second-order spin-orbit interaction. This latter term is usually dominant, as shown, for example, by Dixon for H_2CO [20]. The second-order mechanism arises as follows: If we write the Hamiltonian for rigid-rotation plus spin-orbit coupling as

$$H_1 = A(N_a - L_a)^2 + B(N_b - L_b)^2 + C(N_c - L_a)^2 + \sum_j \eta_j s_j, \quad (2.50)$$

the perturbation is due to the cross term between

$$H_1 = -2AN_aL_a - 2BN_bL_b - 2CN_cL_c \quad (2.51)$$

and

$$H_2 = \sum_j \eta_j s_j, \quad (2.52)$$

where the sum over the j is the sum over all the electrons. Dixon [21] showed that, for a doublet state,

$$\epsilon_{qq} = \sum_{l \neq l_0}\sum_j \frac{4B_q k_j \langle l_0|L_q|l\rangle\langle l|\eta_{qj}|l_0\rangle}{E_l - E_0}, \quad (2.53)$$

where $k_j = +1$ for the promotion of the unpaired electron, -1 for promotion of an electron into a half filled orbital, or zero otherwise. The matrix elements of η_j can usually be represented by using $\eta_j = \zeta_{np} I_j$, where ζ_{np} is the orbital coupling constant derived from atomic spectra. This theory has been applied by Dixon with considerable success, and is similar in philosophy to that derived from the analogous constants in diatomic molecules [22].

2.3.2 Spin-Spin Constants

The contributions to the spin-spin constants, α and β, come from the first-order electron dipole-dipole interaction, and from second-order spin-orbit interaction. The relative size of these contributions has been calculated by Dixon for H_2CO [20]. Similar contributions to the spin-spin splitting constant, λ, in $^3\Sigma$ states of diatomic molecules have been discussed [22].

2.4 Bibliography

[1] G. Duxbury. The electronic spectra of triatomic molecules and the Renner–Teller effect. In R. F. Barrow, D. A. Long, and D. J. Millen, editors, *Molecular Spectroscopy*, volume 3, pages 497–573. The Royal Society of Chemistry, 1975.

[2] J. H. Van Vleck. The coupling of angular momentum vectors in molecules. *Reviews of Modern Physics*, 23:213–227, July 1951.

[3] J. T. Hougen. *The Calculation of Rotational Energy Levels and Rotational Line Intensities in Diatomic Molecules*. NBS monograph. U.S. Government Printing Office, 1970.

[4] J. T. Hougen. Rotational energy levels of a linear triatomic molecule in a $^2\Pi$ electronic state. *The Journal of Chemical Physics*, 36(2):519–534, 1962.

[5] J. W. C. Johns. k-type doubling of linear molecules in $^1\Pi$ electronic states. *Journal of Molecular Spectroscopy*, 15(4):473–482, 1965.

[6] W. Moffitt and A. D. Liehr. Configurational instability of degenerate electronic states. *Physical Review*, 106:1195–1200, June 1957.

[7] J. K. G. Watson. Determination of centrifugal distortion coefficients of asymmetric-top molecules. *The Journal of Chemical Physics*, 46(5):1935–1949, 1967.

[8] J. K. G. Watson. Determination of centrifugal distortion coefficients of asymmetric-top molecules. III. Sextic coefficients. *The Journal of Chemical Physics*, 48(10):4517–4524, 1968.

[9] W. T. Raynes. Spin splittings and rotational structure of nonlinear molecules in doublet and triplet electronic states. *The Journal of Chemical Physics*, 41(10):3020–3032, 1964.

[10] I. C. Bowater, J. M. Brown, and A. Carrington. Microwave Spectroscopy of nonlinear free radicals. 1. General Theory and Application to Zeeman Effect in HCO. *Proceedings of the Royal Society of London Series*

A-Mathematical Physical and Engineering Sciences, 333(1594):265–288, 1973.

[11] F. Creutzberg and J. T. Hougen. Triplet-state rotational energy levels for near-symmetric rotor molecules of symmetry C_{2v}, D_2, and D_{2h}. *Canadian Journal of Physics*, 45(3):1363–1387, 1967.

[12] C. di Lauro. Rotational structure of singlet-triplet transitions in orthorhombic molecules. *Journal of Molecular Spectroscopy*, 35(3):461–475, 1970.

[13] C. C. Lin. Theory of the fine structure of the microwave spectrum of NO_2. *Physical Review*, 116:903–910, November 1959.

[14] R. N. Dixon and G. Duxbury. Doublet splittings in a non-rigid molecule. *Chemical Physics Letters*, 1(8):330–332, 1967.

[15] G. Duxbury. The asymmetric rotor in electronic spectroscopy: Centrifugal distortion and electron spin coupling effects in the spectra of NH_2 and CH_2. *Journal of Molecular Spectroscopy*, 25(1):1–11, 1968.

[16] C. M. Woodman. The absorption spectrum of HNF in the region 3800–5000 Å. *Journal of Molecular Spectroscopy*, 33(2):311–344, 1970.

[17] E. Hill and J. H. Van Vleck. On the quantum mechanics of the rotational distortion of multiplets in molecular spectra. *Physical Review*, 32:250–272, August 1928.

[18] C. di Lauro. Spin-allowed and forbidden rotational intensities in doublet-doublet transitions in asymmetric top molecules. *Journal of Molecular Spectroscopy*, 51(2):356–362, 1974.

[19] R. N. Dixon, G. Duxbury, and H. M. Lamberton. The analysis of a $^2A_1 - {}^2B_1$ electronic band system of the AsH_2 and AsD_2 radicals. *Proceedings of the Royal Society of London. Series A. Mathematical and Physical Sciences*, 305(1481):271–290, 1968.

[20] R. N. Dixon. Electron spin coupling constants for the a, $^3A''$ state of formaldehyde. *Molecular Physics*, 13(1):77–81, 1967.

[21] R. N. Dixon. Spin-rotation interaction constants for bent AH_2 molecules in doublet electronic states. *Molecular Physics*, 10(1):1–6, 1965.

[22] M. Horani, J. Rostas, and H. Lefebvre-Brion. Fine structure of $^3\Sigma^-$ and $^3\Pi$ states of NH, OH^+, PH, and SH^+. *Canadian Journal of Physics*, 45(10):3319–3331, 1967.

CHAPTER 3

The Renner-Teller Effect

CONTENTS

3.1 EFFECTIVE BENDING HAMILTONIANS

Within the adiabatic approach of separation of nuclear and electronic motions, the total wavefunction is expanded in terms of electronic functions $\Psi_{el}^{(i)}$ that depend parametrically on the internal coordinates of the nuclei, and the corresponding nuclear functions $\Psi_{nucl}^{(i)}$,

$$\Psi = \sum_i \Psi_{nucl}^{(i)} \Psi_{el}^{(i)}. \tag{3.1}$$

In the adiabatic or Born-Oppenheimer approximations, only one electronic state is retained. $i = (\Sigma, |\Lambda|, \Gamma)$ are the electronic quantum numbers, with Σ and Λ the a-axis projections of $\boldsymbol{S}$ and $\boldsymbol{L}$, respectively, and Γ a symmetry index. The nuclear wave function corresponding to the electronic state i is approximated as a product of bending, stretching, and rotational functions,

$$\Psi_{nucl}^{(i)} = \Phi_{v_2}^{b,(i)}(\rho)\chi_{v_1}^{s,(i)}(s_s)\chi_{v_3}^{a,(i)}(s_a)D_{mk}^{N}(\theta, \chi, \varphi), \tag{3.2}$$

where $\Phi_{v_2}^{b,(i)}(\rho)$ denotes the bending functions, and $\chi_{v_1}^{s,(i)}(s_s)$ and $\chi_{v_3}^{a,(i)}(s_a)$ the symmetric and antisymmetric stretching functions, respectively. $D_{mk}^{N}(\theta, \chi, \varphi)$ are the Wigner rotational functions, with (θ, χ, φ) the Euler angles describing the orientation of the molecular plane in the laboratory system. The a-axis rotational part of $D_{mk}^{N}(\theta, \chi, \varphi)$ is $(2\pi)^{-1/2}e^{ik\varphi}$. The electronic basis functions may be written as

$$\Psi_{el}^{(\Sigma,\Lambda)} = f^{(\Lambda)}(\varphi - \nu)\phi^{|\Lambda|}(r_1, \ldots r_{3n-1}; \rho, s_s, s_a)|S\Sigma\rangle. \tag{3.3}$$

$(\varphi - \nu)$ is the electronic azimuthal angle with respect to the molecular plane, and $r_1, \ldots r_{3n-1}$ are the remaining electronic coordinates. In the linear molecule limit, the functions $f^{(\Lambda)}(\varphi - \nu)$ take the analytical form

$$f^{(\Lambda)}(\varphi - \nu) = \frac{1}{\sqrt{2\pi}} e^{i\Lambda(\varphi - \nu)}. \tag{3.4}$$

Alternatively, the electronic basis functions may be chosen as

$$\Psi_{el}^{(\Sigma, |\Lambda|, \pm)} = f^{(|\Lambda|, \pm)}(\varphi - \nu)\phi^{|\Lambda|}(r_1, \dots r_{3n-1}; \rho, s_s, s_a)|S\Sigma\rangle, \tag{3.5}$$

where $f^{(|\Lambda|, \pm)}(\varphi - \nu)$ behave as

$$f^{(|\Lambda|, \pm)}(\varphi - \nu) = \frac{1}{\sqrt{4\pi}} \left(e^{i\Lambda(\varphi - \nu)} \pm e^{-i\Lambda(\varphi - \nu)} \right) \tag{3.6}$$

in the linear molecule limit. The difference between the two sets of electronic basis functions is that the former describes free electronic rotation around the linear molecule axis in the two directions, while the latter describes a standing wave.

The Renner–Teller effect is an important example of a so-called Born-Oppenheimer breakdown, for which the one-state approximation of the sum in Equation (3.1) is not valid. Consider a molecule in linear configuration in a degenerate electronic state $|\Lambda| = 1, 2 \dots$. As the molecule bends, the degeneracy is lifted and the electronic state is split into two components, or "half-states," denoted in the following text as upper (u) and lower (l) for simplicity[1]. Hence, the total wavefunction must include a linear combination of two products of electronic and nuclear functions,

$$\Psi = \Psi_{nucl}^{(u)}\Psi_{el}^{(u)} + \Psi_{nucl}^{(l)}\Psi_{el}^{(l)}. \tag{3.7}$$

The most important nonadiabatic interactions arise from the excitation of bending vibrations, and we will initially restrict the discussion to the single bending vibration of a triatomic molecule, with bending coordinate, ρ, the supplement to the interbond angle. This will be followed by studying the effects of vibrational resonances on Renner–Teller coupling in triatomic molecules.

Substitution of Equation (3.7) into the Schrödinger equation, and then integration over all inactive coordinates, then gives rise to two coupled differential equations for the bending functions $\Phi^{b,(u,l)}(\rho)$, for each set of the quantum numbers k, v_1, and v_3

$$\begin{pmatrix} T(\rho) + U^{(u)}(\rho) - E & H^{(u,l)}(\rho) \\ H^{(l,u)}(\rho) & T(\rho) + U^{(l)}(\rho) - E \end{pmatrix} \begin{pmatrix} \Phi^{b,(u)}(\rho) \\ \Phi^{b,(l)}(\rho) \end{pmatrix} = 0. \tag{3.8}$$

The electronic wavefunctions $\Psi_{el}^{(u,l)}$ and this Hamiltonian for Renner–Teller

---

[1] Jungen and Merer [1] use the notation plus $(+)$ and minus $(-)$, while Barrow et al. [2] use the notation (a) and (s), referring to antisymmetric or symmetric behaviour of the electronic wavefunction with respect to reflection in the nuclear plane.

and also spin-orbit coupling, may be chosen either in the context of a perturbed linear molecule, Equations (3.3) and (3.4), or in the context of the coupling between two Born-Oppenheimer states of a bent molecule, Equations (3.5) and (3.6). The linear molecule approach is used by Jungen and Merer [1, 3], while Barrow, Dixon, and Duxbury [2] started from the bent molecule approach. In principle, these two approaches should lead to identical eigenvalues, but, in general, different approximations and numerical truncation errors may be made in the two cases.

The main difference between the linear and bent molecule approaches to Equation (3.8) is the nature of the coupling term $H^{(u,l)}(\rho)$. In the linear molecule approach $\Psi_{el}^{(u,l)}$ are eigenfunctions of the axial orbital angular momentum operator, and the coupling function is electrostatic

$$H^{(u,l)}(\rho) = V^{(u)}(\rho) - V^{(l)}(\rho). \tag{3.9}$$

In contrast, with the bent molecule approach, the coupling function is the sum of the a-axis Coriolis and spin-orbit interactions

$$H^{(u,l)}(\rho) = -2A(\rho)\langle (J_a - S_a)L_a \rangle + \zeta \langle L_a S_a \rangle. \tag{3.10}$$

The effective potential functions in the linear case are

$$U^{(u,l)}(\rho) = \frac{1}{2}\left[V^{(u)}(\rho) + V^{(l)}(\rho)\right] + A(\rho)\langle (J_a - S_a) \pm L_a \rangle^2 \pm \zeta \langle L_a S_a \rangle, \tag{3.11}$$

whereas in the bent molecule approach,

$$U^{(u,l)}(\rho) = V^{(u,l)}(\rho) + A(\rho)\langle (J_a - S_a)^2 + L_a^2 \rangle. \tag{3.12}$$

Unfortunately, neither of these two model Hamiltonians provides a satisfactory approach for molecules, such as CH_2 or NH_2, which at many energies can execute a large amplitude motion from linear to strongly bent configurations. In a linear geometry, the Coriolis coupling (3.10) goes to infinity except for $K = 0$, and when strongly bent, the electrostatic coupling (3.9) becomes far larger than the vibrational intervals. Two methods were proposed to overcome this difficulty. In the first of these, due to Barrow, Dixon, and Duxbury [2] (hereafter denoted BDD), the coupled equations are subjected to a contact transformation chosen to diagonalize all but the nuclear kinetic energy operator in (3.8). This transformation is equivalent to the use of rotationally adiabatic potential curves at each value of ρ.

The second method due to Jungen and Merer [3] (hereafter denoted JM), is based upon the matrix approach of Renner [4, 5]. It used linear molecule angular momentum functions, and retains the two-part form for the zero-order wavefunctions Equation (3.7), but derives two sets of there from the two Born-Oppenheimer potentials. In either method the terms omitted in setting up the zero-order basis then give rise to coupling matrix elements.

There is one fundamental difference between the two approaches. In linear molecules, there is a central component of the manifold of vibronic levels for each value of the bending vibrational quantum number v_2 which has $K = v_2 + \Lambda$. These levels are equally associated with both Born-Oppenheimer potential curves of the two component states, and have been called unique levels by Jungen and Merer [3]. Their linear molecule approach specifically includes basis functions appropriate to the unique levels. In contrast, there are no such functions separately identifiable in the complete set of basis functions from the BDD transformation. It is, therefore, to be expected that for molecules having linear equilibrium configurations for both components of the degenerate state, the JM basis set will provide a much better starting point than the BDD basis set for a numerical diagonalisation of the full Hamiltonian.

However, for molecules with large amplitude motion, the JM method is computationally more demanding than the BDD method. Furthermore, the correlation diagram for NH_2 constructed by Dressler and Ramsay [6] shows that there are no clearly identifiable unique levels in this case. A detailed comparison has been made between calculations using both methods, including a study of the effect of approximating the solutions of the full coupled equations.

In a rotating coordinate system the ro-vibronic Hamiltonian for an XYZ molecule in an orbitally degenerate state can be partitioned in the following way [1, 7], as shown in the derivation of Equation (3.8):

$$H = H_b + H_{str} + H_{ev} + H_{so} + H_r^{b,c}. \tag{3.13}$$

H_b is the large amplitude Hamiltonian of BDD [2] and includes the a-axis rotation

$$H_b = T(\rho) + V^{(u,l)}(\rho) + H_r^a. \tag{3.14}$$

$T(\rho)$ is the nuclear kinetic energy operator, and $V^{(u,l)}(\rho)$ are cuts through the Born-Oppenheimer potential surfaces along the bending coordinate for the two component states. H_{str} is the Hamiltonian for the symmetric and antisymmetric stretching vibrations, which will be dealt with in Section 3.2. The vibronic interaction Hamiltonian is

$$H_{ev} = A(\rho) \left[L_\alpha^2 - 2L_\alpha \left(J_\alpha - S_\alpha \right) \right]. \tag{3.15}$$

The first term is diagonal, and the second off-diagonal in the electronic basis, as written in Equations (3.12) and (3.10). In multiplet states, the a-axis spin-orbit coupling operator $H_{so} = \zeta L_\alpha S_\alpha$ has nonzero matrix elements and must be included.

The first step in the separation of the Hamiltonian is to neglect the b and c-axis components of the overall rotation $H_r^{b,c}$ in zero order. $J_\alpha - S_\alpha$ can then be replaced by the good quantum number K. Similarly, in a strictly linear geometry, L_α can be associated with the quantum number Λ. Departure from this simple form for bent geometries can be simply parameterized, if necessary [1, 2, 8]. Within the bent molecule approach, the coupled Equations (3.2) now take the form

$$\left[\begin{pmatrix} T(\rho) + V^{(u)}(\rho) + A(\rho)\left\{K^2 + \Lambda^2\right\} & -2A(\rho)K\Lambda + \zeta\Lambda\Sigma \\ -2A(\rho)K\Lambda + \zeta\Lambda\Sigma & T(\rho) + V^{(l)}(\rho) + A(\rho)\left\{K^2 + \Lambda^2\right\} \end{pmatrix} - \begin{pmatrix} E & 0 \\ 0 & E \end{pmatrix}\right] \begin{pmatrix} \Phi^{b,(u)}(\rho) \\ \Phi^{b,(l)}(\rho) \end{pmatrix} = 0, \tag{3.16}$$

where

$$A(\rho) = \frac{\hbar^2}{2r_0^2}\left[\frac{m_x + 2m_y}{m_x m_y\left(1 - \cos\rho\right)}\right] \tag{3.17}$$

and

$$T(\rho) = -g(\rho)\frac{\partial^2}{\partial\rho^2} - \frac{\partial^2}{\partial\rho^2}g(\rho) - \frac{1}{2}g(\rho)\left[1 - \frac{3 - \cos^2\rho}{\sin^2\rho}\right], \tag{3.18}$$

with

$$g(\rho) = \frac{\hbar^2}{2r_0^2}\left[\frac{m_x + 2m_y}{m_y\left[m_x + m_y\left(1 - \cos\rho\right)\right]}\right]. \tag{3.19}$$

The third term in $T(\rho)$, (3.18), arises from the noncommutation of the vibrational kinetic energy operator and the inverse inertial matrix μ, as shown by Hougen, Bunker, and Johns [9]. The next step is to transform Equation (3.16) in order to minimize the coupling. The angle $\gamma(\rho)$ of the transformation matrix

$$\mathbf{U}(\rho) = \begin{pmatrix} \cos\gamma(\rho) & \sin\gamma(\rho) \\ -\sin\gamma(\rho) & \cos\gamma(\rho) \end{pmatrix} \tag{3.20}$$

is chosen to diagonalize the potential energy matrix. It takes the value $\gamma(0) = \pi/4$ for a linear conformation and goes to zero when the molecule is strongly bent. The new Hamiltonian

$$\mathbf{H}'' = \begin{pmatrix} T(\rho) + U^{(u)}(\rho) + T_{ev}^{(u,u)}(\rho,\gamma) & T_{ev}^{(u,l)}(\rho,\gamma) \\ T_{ev}^{(l,u)}(\rho,\gamma) & T(\rho) + U^{(l)}(\rho) + T_{ev}^{(l,l)}(\rho,\gamma) \end{pmatrix} \tag{3.21}$$

now has K-dependent adiabatic effective potential curves

$$\begin{aligned} U^{(i)}(\rho) &= \frac{1}{2}\left\{V^{(u)}(\rho) + V^{(l)}(\rho)\right\} + A(\rho)\left\{K^2 + \Lambda^2\right\} \\ &\pm \left[\frac{1}{4}\left\{V^{(u)}(\rho) - V^{(l)}(\rho)\right\}^2 + \left\{2A(\rho)K - \zeta\Sigma\right\}^2\Lambda^2\right]^{\frac{1}{2}} \end{aligned} \tag{3.22}$$

and contains small matrix elements T_{ev}, which are generated due to the noncommutation of the transformation matrix with the kinetic energy operator

$$\mathbf{T}_{ev} = \begin{pmatrix} 2g^{\rho\rho}(\rho)\left(\frac{\partial\gamma(\rho)}{\partial\rho}\right)^2 & [T(\rho),\gamma(\rho)] \\ [\gamma(\rho),T(\rho)] & 2g^{\rho\rho}(\rho)\left(\frac{\partial\gamma(\rho)}{\partial\rho}\right)^2 \end{pmatrix}. \tag{3.23}$$

Having minimized the vibronic coupling in this way, the off-diagonal operator in (3.21) can be dropped in zero-order, thereby separating the coupled equations into two equations for the vibrational wavefunctions $\Phi^{b,(u)}(\rho)$ and $\Phi^{b,(l)}(\rho)$ for the upper and lower-component states, respectively.

The equations for large amplitude motion involving the potential curves, which are radially adiabatic at each value of ρ, can be written as, denoting again the bending functions as $\Phi^{b,(i)}(\rho)$,

$$\left[T(\rho) + T_{ev}(\rho)^{(i,i)} + U^{(i)}(\rho) - E \right] \Phi^{b,(i)}(\rho) = 0. \tag{3.24}$$

The solution of (3.24) can be effected by numerical integration, as described originally by Hougen, Bunker, and Johns [9] for the nondegenerate case. The method we describe is based upon Cooley's adaptation [10] of the Numerov method [11, 12].

The differential Equation (3.24) is of the general form

$$g(\rho)\frac{d^2\Phi(\rho)}{d\rho^2} + \frac{d^2}{d\rho^2}\left[g(\rho)\Phi(\rho)\right] = \left[U(\rho) - E\right]\Phi(\rho), \tag{3.25}$$

where $g(\rho)$ is a slowly varying function. By expanding $g(\rho)$ and $\Phi(\rho)$ as Taylor series, and applying the Numerov correction, Equation (3.25) can be converted into a difference equation. Let $\rho_i = ih, i = 0, 1, 2, \ldots, n+1$ and $P_i = \Phi(\rho_i)$, then (3.25) becomes

$$G_{i,i-1}P_{i-1} + G_{ii}P_i + G_{i,i+1}P_{i+1} = 0, \tag{3.26}$$

where

$$G_{ii} = 4g_i + \frac{2}{3}h^2\left(U_i - E\right) - \frac{h^4}{24}\frac{(U_i - E)^2}{g_i} \tag{3.27}$$

and

$$G_{i-1,i} = -\left(g_{i-1} + g_i\right) - \frac{h^2}{12}\left(U_{i-1} + U_i - 2E\right). \tag{3.28}$$

$U(\rho)$ includes a pole at $\rho = 0$, and if it is expanded about $\rho = 0$ as a polynomial in ρ, the leading term is found to be

$$U(\rho) = \frac{2\left(l^2 - \frac{1}{4}\right)}{\rho^2}g(\rho) + \cdots . \tag{3.29}$$

l is integral, even in the case that we are now considering, which includes partial Renner uncoupling, and takes the value $K \pm \Lambda$ for $\Phi^{(u)}$ and $\Phi^{(l)}$. This pole imposes a boundary condition on $\Phi(\rho)$, such that it can be expanded as

$$\Phi(\rho) = \rho^{l+\frac{1}{2}}\left[a_0 + a_2\rho^2 + a_4\rho^4 + \ldots\right] \equiv \rho^{l+\frac{1}{2}}\chi(\rho). \tag{3.30}$$

Thus, the standard Cooley-Numerov method will fail at low values of ρ, since

$\rho^{l+\frac{1}{2}}$ cannot be expanded as a convergent polynomial in integer powers of ρ. However, if we let

$$W(\rho) = U(\rho) - \frac{2\left(l^2 - \frac{1}{4}\right)}{\rho^2} g(\rho), \tag{3.31}$$

then (3.25) becomes

$$2g\frac{d^2\chi}{d\rho^2} + 2\left[\frac{2l+1}{\rho} + \frac{dg}{d\rho}\right]\frac{d\chi}{d\rho} = \left[W - \frac{(2l+1)}{\rho}\frac{dg}{d\rho} - \frac{d^2g}{d\rho^2} - E\right]\chi. \tag{3.32}$$

Since both g and χ are even functions of ρ, the derivatives $(dg/d\rho)_0$ and $(d\chi/d\rho)_0$ are both zero, so that there are no poles in (3.32). If we assume that the variation of g with ρ can be neglected as $\rho \to 0$, and that $W(\rho)$ is expanded as a power series in the same region,

$$W(\rho) = W_0 + W_2\rho^2 + W_4\rho^4 + +, \tag{3.33}$$

then it is possible to obtain explicit relationships between the coefficients of χ and those of W:

$$\frac{a_2}{a_0} = \frac{W_0 - E}{8(l+1)g},$$

$$\frac{a_4}{a_0} = \frac{(W_0 - E)^2}{128(l+1)(l+2)g^2} + \frac{W_2}{16(l+2)g}, \tag{3.34}$$

$$\frac{a_6}{a_0} = \frac{(W_0 - E)^3}{3072(l+1)(l+2)(l+3)g^3} + \frac{(W_0 - E)W_2(3l+5)}{384(l+1)(l+2)(l+3)g^2} + \frac{W_4}{24(l+3)g}.$$

The first n parameters W_j can be calculated from the first n computed values of $W(\rho_i)$, allowing the series solution to Equation (3.17) to be derived analytically. The solution of Equation (3.26) can be obtained by methods analogous to those for diagonalising matrices of bandwidth three, such as the Sturm sequence algorithm [13, 14].

In seeking the eigenvalues E_v by such node counting methods, it is either necessary to start at $\rho = 0$, where $\chi = a_0$, or to chose some ρ_0 and correctly reduce the three terms in (3.26) to two terms for the first difference equation

$$G'_{i_0,i_0}P_{i_0} + G_{i_0,i_0+1}P_{i_0+1} = 0. \tag{3.35}$$

This can be achieved in the following manner. First, calculate χ_{i_0-1} and χ_{i_0} from Equation (3.26). Hence,

$$P_{i_0-1} = (i_0 - 1)^{l+\frac{1}{2}}\chi_{i_0-1}$$
$$P_{i_0} = i_0^{l+\frac{1}{2}}\chi_{i_0} \tag{3.36}$$

apart from an unknown scale factor. Then, writing

$$P_{i_0-1} = \left(\frac{i_0-1}{i_0}\right)^{l+\frac{1}{2}}\left(\frac{\chi_{i_0-1}}{\chi_{i_0}}\right)P_{i_0}, \tag{3.37}$$

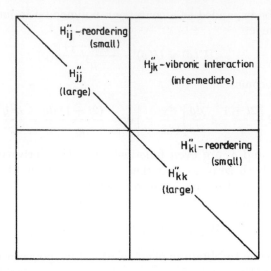

Figure 3.1: General form of the vibronic matrix at the H'' (JM) level. In the BDD model, the reordering elements are identically zero. From Duxbury and Dixon, *Mol. Phys.* 43:255, 1981 [15], reprinted with permission of the publisher (Taylor & Francis Ltd).

substitution into (3.26) gives

$$\left[G_{i_0,i_0} + G_{i_0,i_0-1} \left(\frac{i_0 - 1}{i_0} \right)^{l+\frac{1}{2}} \left(\frac{\chi_{i_0-1}}{\chi_{i_0}} \right) \right] P_{i_0} + G_{i_0,i_0+1} P_{i_0+1} = 0. \quad (3.38)$$

In general, the use of $i_0 = 10$ ensures that the solutions are free from any errors arising from the singularity at the origin in (3.25). They are then used as zero-order basis to set up the hamiltonian matrix, $\mathbf{H}''$, of (3.21). Diagonalization yields the eigenfunctions of (3.16). The form of the $\mathbf{H}''$ matrix is shown in Figure 3.1. The matrix elements of the off-diagonal block, coupling the two electronic states, are now rather small.

In our original approach, we chose to use a bending potential based upon a harmonic potential perturbed by a Lorentzian hump

$$V(\rho) = \frac{1}{2} f \rho^2 + \frac{a}{1 + b\rho^2} - V_m. \quad (3.39)$$

For a bent state, this may be conveniently expressed [2] in terms of a barrier height H, the equilibrium angle ρ_m, and the quadratic force constant f_m at the minimum in V:

$$V(\rho) = \frac{H f_m \left(\rho^2 - \rho_m^2 \right)^2}{f_m \rho_m^4 + (8H - f_m \rho_m^2) \rho^2}, \quad (3.40)$$

where the equivalences between the parameters are

$$\rho_m^2 = \frac{1}{B}\left[\left(\frac{2ab}{f}\right)^{\frac{1}{2}} - 1\right] \tag{3.41}$$

$$H = a\left[1 - \left(\frac{f}{2ab}\right)^{\frac{1}{2}}\right]^2 \tag{3.42}$$

$$f_m = 4f\left[1 - \left(\frac{f}{2ab}\right)^{\frac{1}{2}}\right] \tag{3.43}$$

and

$$V_m = a\left[2\left(\frac{f}{2ab}\right)^{\frac{1}{2}} - \frac{f}{2ab}\right]. \tag{3.44}$$

In many cases, it is necessary to add further parameters to increase the flexibility of the potential, while retaining this basic three parameter form [2].

3.2   INCLUSION OF STRETCHING VIBRATIONS: THE STRETCH-BENDER APPROACH

The methods presented above efficiently treat the bending vibration and its coupling with a-axis rotation due to the Renner–Teller effect. An extension, named stretch-bender approach [16], has been developed to include the symmetric and antisymmetric stretch vibrations. In essence, for each set of stretch quantum numbers, v_1 and v_3, a Renner–Teller problem is solved to yield zero-order bending functions. An interaction matrix is then set up to treat stretch-stretch interactions.

To derive the stretch-bender Hamiltonian, the cartesian coordinates of the three atoms are expressed in terms of the coordinates of a reference configuration and displacements. The reference configuration is chosen, such as to follow the minimum path on the potential surface as the molecule bends. Within this model, the kinetic energy operator can be written in terms of the bending coordinate of the reference frame, ρ, the symmetric and antisymmetric stretch coordinates, s^s and s^a, and the angle φ corresponding to a rotation around the molecular a-axis.

$$T = -\frac{\hbar^2}{2}\left[-g^{\rho\rho}f(\rho) + g^{\rho\rho}\frac{\partial^2}{\partial\rho^2}\right] - \frac{\hbar^2}{2}g^{ss}\frac{\partial^2}{\partial s^{s2}} - \frac{\hbar^2}{2}g^{aa}\frac{\partial^2}{\partial s^{a2}} - \frac{\hbar^2}{2}g^{\varphi\varphi}\frac{\partial^2}{\partial\varphi^2}, \tag{3.45}$$

where

$$f(\rho) = -\frac{1 + \sin^2(\rho/2)}{16\sin^2(\rho/2)}. \tag{3.46}$$

$\frac{\hbar^2}{2}g^{\varphi\varphi} \equiv A$ is the a-axis rotational constant. The potential energy surface $V_{inst}(\bar{\rho}, r_{12}, r_{13})$, as it can be obtained from *ab initio* calculations, is, however, a function of the instantaneous coordinates. For the use within the stretch-bender approach, it is, therefore, expanded around the reference configuration, which yields in good approximation [16]

$$V_{inst}(\bar{\rho}, r_{12}, r_{13}) \approx V^b(\rho, s^s) + V^{as}(\rho, s^s, s^a), \tag{3.47}$$

where the first term relates to bending and the second to stretching. The first term can further be approximated by

$$V^b(\rho, s^s) = V^b(\rho, 0) + \frac{\partial V^b(\rho, s^s)}{\partial s^s} s^s + \cdots , \tag{3.48}$$

where $V^b(\rho, 0)$ is the bending potential at the reference configuration. We have considered two ways of expanding the part of the potential energy function associated with bond stretching, $V^{as}(\rho, s^s, s^a)$. In the first, we assume that the potential is almost harmonic, with a small anharmonic correction. In the second, we assume that the potential can be expanded in terms of Morse potentials for the bonds. With either choice of the potential, the symmetric and antisymmetric stretching functions, $\chi^s(s^s; \rho)$ and $\chi^a(s^a; \rho)$, are obtained analytically. Different from the functions used in (3.2), they depend parametrically on the bending coordinate ρ. For a given electronic state i, the nuclear wavefunction can now be written as

$$\Psi_{nucl}^{(i)} = \Phi_{v_1, v_2, v_3}^{b,(i)}(\rho) \chi_{v_1}^{s,(i)}(s_s; \rho) \chi_{v_3}^{a,(i)}(s_a; \rho) D_{mk}^N(\theta, \chi, \varphi). \tag{3.49}$$

After integrating over all coordinates except of ρ, the effective bending equations within the stretch-bender approach become

$$\left[\left\langle \chi_{v_1'}^s \chi_{v_3'}^a | T(\rho) + U(\rho) | \chi_{v_3}^a \chi_{v_1}^s \right\rangle - \delta_{v_1' v_1} \delta_{v_3' v_3} E_{v_1, v_2, v_3} \right] \Phi_{v_1, v_2, v_3}(\rho) = 0. \tag{3.50}$$

In contrast to the large amplitude bender model, where a single bending equation, (3.24), is obtained, and where the bending function does not depend on the stretching state, (3.50) consists of a system of coupled equations. For each set of v_1 and v_3, one particular bending function is obtained, as indicated by the notation $\Phi_{v_1, v_2, v_3}(\rho)$. The matrix elements of the operators $T(\rho)$ and $U(\rho)$ in (3.50) are

$$
\begin{aligned}
T_{v_1' v_3' v_1 v_3}(\rho) =\ & -\frac{\hbar^2}{2} g^{\rho\rho} \left[-f(\rho) + \frac{\partial^2}{\partial \rho^2} \right] \delta_{v_1' v_1} \delta_{v_3' v_3} \\
& -\frac{\hbar^2}{2} g^{\rho\rho} \left[I_{v_1' v_1}^{(2)} + I_{v_3' v_3}^{(2)} + \left(I_{v_1' v_1}^{(1)} + I_{v_3' v_3}^{(1)} \right) \frac{\partial}{\partial \rho} + 2 I_{v_1' v_1}^{(1)} I_{v_3' v_3}^{(1)} \right]
\end{aligned}
\tag{3.51}
$$

and

$$U_{v_1' v_3' v_1 v_3}(\rho) = \left\{ V^b(\rho, 0) + \frac{\hbar^2}{2} g^{\varphi \varphi} K^2 \right.$$

$$+ \hbar \omega_1(\rho) \left(v_1 + \frac{1}{2} \right) \left[1 + x_1 \left(v_1 + \frac{1}{2} \right) \right]$$

$$\left. + \hbar \omega_3(\rho) \left(v_3 + \frac{1}{2} \right) \left[1 + x_3 \left(v_3 + \frac{1}{2} \right) \right] \right\} \delta_{v_1' v_1} \delta_{v_3' v_3}$$

$$+ \frac{\partial V_0}{\partial \rho} \frac{\partial \rho}{\partial s^s} F_{v_1' v_1}. \tag{3.52}$$

Three integrals, $I^{(1)}$, $I^{(2)}$, and F, appear in the above definitions. They are the off-diagonal kinetic energy correction terms

$$I_{v_i' v_i}^{(1)}(\rho) = \int_{-\infty}^{\infty} \chi_{v_i'}^* \frac{\partial \chi_{v_i}}{\partial \rho} ds^i, \tag{3.53}$$

the diagonal and off-diagonal anharmonic correction terms

$$I_{v_i' v_i}^{(2)}(\rho) = \int_{-\infty}^{\infty} \chi_{v_i'}^* \frac{\partial^2 \chi_{v_i}}{\partial \rho^2} ds^i, \tag{3.54}$$

and the Fermi resonance integral $F_{v_i' v_i}(\rho)$

$$F_{v_i' v_i}(\rho) = \int_{-\infty}^{\infty} \chi_{v_i'}^* s^i \chi_{v_i} ds^i. \tag{3.55}$$

v_i are either v_1 or v_3. If the stretching functions are approximated as harmonic oscillator functions, the integrals take the explicit form given in Table 3.1.

Table 3.1: Form of the Perturbation Integrals Arising In the Stretch-Bender Approach When Evaluated In the Harmonic Oscillator Approximation

$$
\begin{aligned}
I^{(1)}_{v'_i v_i}(\rho) &= \int_{-\infty}^{\infty} \chi^*_{v'_i} \frac{\partial \chi_{v_i}}{\partial \rho} ds^i \\
&= \frac{1}{2\alpha_i} \frac{d\alpha_i}{d\rho} \left(-\delta_{v'_i v_i +2} \sqrt{(v_i+1)(v_i+2)} + \delta_{v'_i v_i -2} \sqrt{v_i(v_i-1)} \right) \\
I^{(2)}_{v'_i v_i}(\rho) &= \int_{-\infty}^{\infty} \chi^*_{v'_i} \frac{\partial^2 \chi_{v_i}}{\partial \rho^2} ds^i \\
&= \delta_{v'_i v_i +4} \frac{1}{4} \left(\frac{d\alpha_i}{d\rho}\right)^2 \sqrt{(v_i+1)(v_i+2)(v_i+3)(v_i+4)} \\
&\quad -\delta_{v'_i v_i +2} \frac{1}{2} \left[\frac{d^2\alpha_i}{d\rho^2} - \left(\frac{d\alpha_i}{d\rho}\right)^2 \right] \sqrt{(v_i+1)(v_i+2)} \\
&\quad +\delta_{v'_i v_i} \frac{1}{2} \left(\frac{d\alpha_i}{d\rho}\right)^2 \left(v_1^2 + v_i + 1\right) \\
&\quad +\delta_{v'_i v_i -2} \frac{1}{2} \left[\frac{d^2\alpha_i}{d\rho^2} - \left(\frac{d\alpha_i}{d\rho}\right)^2 \right] \sqrt{(v_i-1)v_i} \\
&\quad +\delta_{v'_i v_i -4} \frac{1}{4} \left(\frac{d\alpha_i}{d\rho}\right)^2 \sqrt{(v_i-3)(v_i-2)(v_i-1)v_i} \\
F_{v'_i v_i}(\rho) &= \int_{-\infty}^{\infty} \chi^*_{v'_i} s^i \chi_{v_i} ds^i \\
&= \delta_{v'_i v_i +1} \frac{1}{\alpha_i} \sqrt{\frac{v_i+1}{2}} + \delta_{v'_i v_i -1} \frac{1}{\alpha_i} \sqrt{\frac{v_i}{2}}
\end{aligned}
$$

The g-functions and force constants and, hence, the stretching frequencies become ρ-dependent. The stretching functions can be written as $\chi_{v_i}(s^i;\rho) = \sqrt{\alpha(\rho)/\pi^{1/2} 2^{v_i} v_i!}\, e^{-y^2/2} H_{v_i}(y)$, where $H(y)$ are Hermite polynomials and $y = \alpha(\rho)s^i$, $\alpha^2(\rho) = \frac{\omega(\rho)}{\hbar g^{ii}(\rho)}$.

To solve the bending Equations (3.50), the integrals $I^{(1)}$, $I^{(2)}$, and F are treated as perturbations. They are neglected in the first step to yield zero-order bending functions, $\Phi^0_{v_1,v_2,v_3}(\rho)$. The full Hamiltonian is then diagonalized in this basis. The treatment can be carried out at four levels, permitting separation of the effects of the various perturbation terms:

1. Neglect all perturbation terms; this is the semi-rigid bender limit.

2. Include the potential coupling only; we then evaluate

$$
\left\langle \phi^0_{v'_1,v'_2,v'_3}(\rho) \left| \frac{\partial V_0}{\partial \rho} \frac{\partial \rho}{\partial s^s} F_{v'_1 v_1} \right| \phi^0_{v_1,v_2,v_3}(\rho) \right\rangle. \tag{3.56}
$$

3. Add the adiabatic correction to the effective potential, evaluating

$$\left\langle \phi^0_{v'_1,v'_2,v'_3}(\rho) \left| -\frac{\hbar^2}{2} g^{\rho\rho} I^{(2)}_{v'_i v_i} \right| \phi^0_{v_1,v_2,v_3}(\rho) \right\rangle, \tag{3.57}$$

where v_i is either v_1 or v_3.

4. Include the nonadiabatic coupling terms

$$\left\langle \phi^0_{v'_1,v'_2,v'_3}(\rho) \left| -\frac{\hbar^2}{2} g^{\rho\rho} I^{(1)}_{v'_i v_i} \frac{\partial}{\partial\rho} \right| \phi^0_{v_1,v_2,v_3}(\rho) \right\rangle, \tag{3.58}$$

where v_i is either v_1 or v_3. The other term arising from the nonadiabatic coupling will be the resonance between stretching states with two quanta of excitation, the Darling-Dennison resonances. This will be calculated using the product of the integrals $I^{(1)}_{v'_1 v_1} I^{(1)}_{v'_3 v_3}$, giving rise to

$$I^{(1)}_{v'_1 v_1}(\rho) I^{(1)}_{v'_3 v_3}(\rho) \sim \delta_{v'_1 v_1+2} \sqrt{(v_1+1)(v_1+2)} \, \delta_{v'_3 v_3-2} \sqrt{v_3(v_3-1)} \tag{3.59}$$

and the equivalent expressions when v_1 and v_3 are interchanged. In addition, there are terms with $\delta_{v_1,v_1\pm2} \delta_{v_3,v_3\pm2}$.

In the approach so far, we have described a model for an isolated electronic state of a molecule, such as water. It can easily be extended for the case of two coupled Renner–Teller states. The two states are approximately decoupled for each combination of v_1 and v_3 and the appropriate zero-order bending functions obtained, just as described in Section 3.1. A large interaction matrix is then set up for fixed values of K and Σ; its structure is shown in Table 3.2. Each diagonal block is of the form shown in Figure 3.1.

3.3 INCLUSION OF OVERALL ROTATION

Of the different terms in the Hamiltonian (3.13), the overall rotation one has not yet been taken into account. The respective term is

$$H^{b,c}_r = BR^2_b + CR^2_c, \tag{3.60}$$

where $B \equiv B(\rho)$ and $C \equiv C(\rho)$ are ρ-dependent rotational "constants" and $\boldsymbol{R}$ the operator for the rotation of the molecular frame introduced by Hougen [17],

$$\boldsymbol{R} = \boldsymbol{J} - \boldsymbol{L} - \boldsymbol{S}. \tag{3.61}$$

Equation (3.60) can be rearranged to give

$$H^{b,c}_r = \frac{1}{2}(B+C)\left(R^2_b + R^2_c\right) + \frac{1}{2}(B-C)\left(R^2_b - R^2_c\right). \tag{3.62}$$

If the definition of $\boldsymbol{R}$ is inserted into the above expression for $H^{b,c}_r$, diagonal terms in the angular momenta involved, as well as various cross terms, are

Table 3.2: Block Structure of the Interaction Matrix (Upper Triangle) for a $^2A_1/^2B_1$ Pair of Electronic States and Given K and Σ, Without "Overall Rotation"

(v_1,v_3)	(0,0)		(1,0)		(2,0)		(0,2)		(1,2)		(2,2)		...
(0,0)	$^2A_1$ RT	FR 0	NA 0	0 0	0 0	DD 0							
	RT $^2B_1$	0 FR	0 NA	0 0	0 0	0 DD							
(1,0)			$^2A_1$ RT	FR 0	0 0	0 0	0 0						
			RT $^2B_1$	0 FR	0 0	0 0	0 0						
(2,0)					$^2A_1$ RT	DD 0	0 0	0 0					
					RT $^2B_1$	0 DD	0 0	0 0					
(0,2)							$^2A_1$ RT	FR 0	NA 0				
							RT $^2B_1$	0 FR	0 NA				
(1,2)									$^2A_1$ RT	FR 0			
									RT $^2B_1$	0 FR			
(2,2)											$^2A_1$ RT		
											RT $^2B_1$		
⋮													

$^2A_1$ and $^2B_1$ denote the diagonal blocks containing the v_2 bending levels. RT = Renner–Teller terms, FR = Fermi Resonance terms, NA = nonadiabatic ($\Delta v = \pm 2$) terms, DD = Darling-Dennison terms.

generated. The former can be evaluated as $J_b^2 + J_c^2 = J^2 - J_a^2$ etc., while evaluation of the latter involves the shift operators

$$J^\pm = J_b \pm iJ_c \qquad (3.63)$$
$$L_\pm = L_b \pm iL_c$$
$$S_\pm = S_b \pm iS_c.$$

The effect of the shift operators acting on a suitable ket is

$$L_\pm|L\Lambda\rangle = f_\pm(L,\Lambda)|L\Lambda \pm 1\rangle \qquad (3.64)$$
$$S_\pm|S\Sigma\rangle = f_\pm(S,\Sigma)|S\Sigma \pm 1\rangle,$$

but

$$J^\pm|Jp\rangle = f_\mp(J,p)|Jp \mp 1\rangle, \qquad (3.65)$$

where

$$f_\pm(j,m) = \hbar\sqrt{j(j+1) - m(m \pm 1)}. \qquad (3.66)$$

The superscript $\pm$ on J indicates the "reversed" angular momentum rules obeyed by J; see also Section 2.1.3. When $H_r^{b,c}$ is expressed in terms of these operators, it takes the form

$$
\begin{aligned}
H_r^{b,c} &= \frac{1}{2}\,(B+C)\left[J^2 - J_a^2 + S^2 - S_a^2\right] \\
&\quad - \frac{1}{2}\,(B+C)\left[J^+ S_- + J^- S_+\right] \\
&\quad + \frac{1}{4}\,(B-C)\left[J^{+2} + J^{-2} + S_+^{\;2} + S_-^2\right] \\
&\quad - \frac{1}{2}\,(B-C)\left[J^+ S_+ + J^- S_-\right].
\end{aligned}
\tag{3.67}
$$

Note that in the above expression, no terms involving the electronic angular momentum L are retained. The diagonal terms contribute to the electronic energy only and can be included in the definition of the potential energy surface, while the off-diagonal terms, which couple different electronic states, are neglected.

The matrix representation of the operator (3.67) is diagonal in the electronic quantum numbers $i = (\Sigma, |\Lambda|, \Gamma)$. It is also diagonal in the quantum numbers v_1 and v_3, since in our model the rotational "constants" A, B, and C do not depend explicitly on the symmetric and antisymmetric stretching coordinates. However, it is not diagonal in v_2 and the rotational quantum numbers k and Σ. Explicit expressions for the matrix elements are readily obtained after integration over all coordinates except the bending coordinate

$$
\begin{aligned}
\langle J\,p\,S\,\Sigma\,&|H_r^{b,c}|\,J\,p'\,S\,\Sigma'\rangle = \\
&\frac{1}{2}\,(B+C)\langle J\,p\,S\,\Sigma\,|J^2 - J_a^2 + S^2 - S_a^2|\,J\,p\,S\,\Sigma\rangle \\
&- \frac{1}{2}\,(B+C)\left[\langle J\,p\,S\,\Sigma\,|J^- S_+|\,J\,p-1\,S\,\Sigma-1\rangle\right. \\
&\qquad\qquad \left.+ \langle J\,p\,S\,\Sigma\,|J^+ S_-|\,J\,p+1\,S\,\Sigma+1\rangle\right] \\
&+ \frac{1}{4}\,(B-C)\left[\langle J\,p\,S\,\Sigma\,|J^{-2}|\,J\,p-2\,S\,\Sigma\rangle + \langle J\,p\,S\,\Sigma\,|J^{+2}|\,J\,p+2\,S\,\Sigma\rangle\right. \\
&\qquad\qquad \left.+ \langle J\,p\,S\,\Sigma\,|S_+^2|\,J\,p\,S\,\Sigma-2\rangle + \langle J\,p\,S\,\Sigma\,|S_-^2|\,J\,p\,S\,\Sigma+2\rangle\right] \\
&- \frac{1}{2}\,(B-C)\left[\langle J\,p\,S\,\Sigma\,|J^+ S_+|\,J\,p+1\,S\,\Sigma-1\rangle\right. \\
&\qquad\qquad \left.+ \langle J\,p\,S\,\Sigma\,|J^- S_-|\,J\,p-1\,S\Sigma+1\rangle\right].
\end{aligned}
\tag{3.68}
$$

Using Equations (3.64) to (3.66) for the shift operators, we finally arrive at the desired expressions

$$
\begin{aligned}
\langle J\,p\,S\,\Sigma\,|H_r^{b,c}|\,J\,p'\,S\,\Sigma'\rangle = & \\
& \frac{1}{2}\,(B+C)\,\big[J(J+1)-p^2+S(S+1)-\Sigma^2\big]\,\delta_{p,p'}\delta_{\Sigma,\Sigma'}\delta_{k,k'} \\
- & \frac{1}{2}\,(B+C)\,[f_+(J,p-1)f_+(S,\Sigma-1)\delta_{p-1,p'}\delta_{\Sigma-1,\Sigma'}\delta_{k,k'} \\
& + f_-(J,p+1)f_-(S,\Sigma+1)\delta_{p+1,p'}\delta_{\Sigma+1,\Sigma'}\delta_{k,k'}] \\
+ & \frac{1}{4}\,(B-C)\,[f_+(J,p-1)f_+(J,p-2)\delta_{p-2,p'}\delta_{\Sigma,\Sigma'}\delta_{k-2,k'} \\
& + f_-(J,p+1)f_-(J,p+2)\delta_{p+2,p'}\delta_{\Sigma,\Sigma'}\delta_{k+2,k'} \\
& + f_+(S,\Sigma-1)f_+(S,\Sigma-2)\delta_{p,p'}\delta_{\Sigma-2,\Sigma'}\delta_{k+2,k'} \\
& + f_-(S,\Sigma+1)f_-(S,\Sigma+2)\delta_{p,p'}\delta_{\Sigma+2,\Sigma'}\delta_{k-2,k'}] \\
- & \frac{1}{2}\,(B-C)\,[f_-(J,p+1)f_+(S,\Sigma-1)\delta_{p+1,p'}\delta_{\Sigma-1,\Sigma'}\delta_{k+2,k'} \\
& + f_+(J,p-1)f_-(S,\Sigma+1)\delta_{p-1,p'}\delta_{\Sigma+1,\Sigma'}\delta_{k-2,k'}]\,. \quad (3.69)
\end{aligned}
$$

The above equations show that the matrix representation of the operator $H_r^{b,c}$ is real and symmetric in all quantum numbers. For the sake of clarity, we have included in Equation (3.69) also the quantum number k, which follows from p and Σ, since

$$
k = p - \Sigma. \tag{3.70}
$$

In Equation (3.69), the matrix elements that are diagonal in Σ, lines 2, 5, and 6, are the usual asymmetric top matrix elements. Those diagonal in k but off-diagonal in Σ, lines 3 and 4, are the spin-uncoupling matrix elements. In addition to these, there are asymmetric rotor-like higher-order terms that are off-diagonal in both k and Σ, lines 7 to 10.

If overall rotation is included, K-blocks are coupled with increments of 2, due to the J^{+2} and J^{-2} operators in Equation (3.67), while the spin components, Σ, are coupled with increments of 1, due to the spin uncoupling operator in the second line of Equation (3.67). There is also a cross term due to the operators in the last line with matrix elements $\langle \Sigma, k\,|\,\Sigma \pm 1, k \mp 2\rangle$. These coupling terms are diagonal in the electronic quantum number Λ. In the final interaction matrix, diagonal blocks of the kind shown in Figure 3.1 are connected through the various rotational coupling terms just discussed. Since the parity of K is conserved within our model, the final interaction matrix contains two separate blocks. Their structures are shown in Tables 3.3 and 3.4 for even and odd values of K, respectively, and $J = \frac{3}{2}$.

Table 3.3: Block Structure of the Full Interaction Matrix for $J = \frac{3}{2}$, $S = \frac{1}{2}$, and K even

(k, Σ)	$p = -\frac{3}{2}$ $(-2, \frac{1}{2})$	$p = -\frac{1}{2}$ $(0, -\frac{1}{2})$	$p = \frac{1}{2}$ $(0, \frac{1}{2})$	$p = \frac{3}{2}$ $(2, -\frac{1}{2})$
$(-2, \frac{1}{2})$		CR	OR	OR
$(0, -\frac{1}{2})$	CR		SU	0
$(0, \frac{1}{2})$	OR	SU		CR
$(2, -\frac{1}{2})$	OR	0	CR	

The diagonal elements are given by the matrix shown in Table 3.2 with the diagonal rotation terms (first line of Equation (3.67)) added. OR = Overall Rotation terms (third line of Equation (3.67)), SU = Spin Uncoupling terms (second line 2 of Equation (3.67)), CR = Cross terms (fourth line of Equation (3.67)).

Table 3.4: Block Structure of the Full Interaction Matrix for $J = \frac{3}{2}$, $S = \frac{1}{2}$, and K odd

(k, Σ)	$p = -\frac{3}{2}$ $(-1, -\frac{1}{2})$	$p = -\frac{1}{2}$ $(-1, \frac{1}{2})$	$p = \frac{1}{2}$ $(1, -\frac{1}{2})$	$p = \frac{3}{2}$ $(1, \frac{1}{2})$
$(-1, -\frac{1}{2})$		SU	OR	CR
$(-1, \frac{1}{2})$	SU		CR	OR
$(1, -\frac{1}{2})$	OR	CR		SU
$(1, \frac{1}{2})$	0	CR	SU	

As Table 3.3, but with K odd.

Inclusion of overall rotation to the stretch-bender approach is described in detail in Reference [18]. The first applications were dedicated to the analysis of the $\tilde{X}^2 B_1$ ground state [19] and the $\tilde{A}^2 A_1$ excited state [20] of NH_2.

3.4 SUMMARY AND ALTERNATIVE FORMULATIONS

In the preceding parts of this chapter, we have presented two methods appropriate for a description of the Renner–Teller effect: One developed by Barrow, Dixon, and Duxbury [2] and the other developed by Jungen and Merer [3]. Both start from effective Hamiltonians that describe the rovibronic coupling due to the bending vibration and a-axis rotation but using different electronic basis functions. The infinite coupling terms are treated by contact transformations, leading to approximately decoupled bending equations and weak remaining nonadiabatic coupling terms. Symmetric and antisymmetric stretching vibrations and b and c-axis rotation are included at a later stage.

Bunker, Jensen, and coworkers [21] have also developed a variational method to treat the Renner–Teller effect. It is a generalization of Jensen's Morse Oscillator Rigid Bender Internal Dynamics (MORBID) approach [22], in which the Hougen, Bunker, and Johns Hamiltonian [9] is expanded in terms of Morse oscillator functions for the stretching functions. Bending basis functions are obtained numerically. The first application was for the methylene molecule [21].

The Renner–Teller effect has also been approached by *ab initio* methods. Pioneering work was initiated by Carter and Handy in the 1980s. They developed a variational Renner–Teller code that uses the exact kinetic energy operator in valence coordinates [23]. Ab initio potential energy surfaces are employed in their calculations, sometimes adjusted to fit experimental data, since at that time ab initio calculations of potential energy surfaces could not be performed to the required accuracy. Important work by Carter, Handy, Rosmus, and coworkers includes the molecules CH_2^+ [23], CH_2 [24], NH_2 [25, 26], MgNC [27], CO_2^+ [28], H_2O^+ [29], and CCO [30]. Extensions of their original code have been made to treat rovibronic coupling between three electronic states [31,32]. Perić and coworkers have developed the theory of the Renner–Teller effect in polyatomic, i.e., tetratomic, pentatomic, and higher, molecules [33–36].

3.5   Bibliography

[1] Ch. Jungen and A. J. Merer. Orbital angular momentum in triatomic molecules. I. A general method for calculating the vibronic energy levels of states that become degenerate in the linear molecule (the Renner–Teller effect). *Molecular Physics*, 40(1):1–23, 1980.

[2] T. Barrow, R. N. Dixon, and G. Duxbury. The Renner effect in a bent triatomic molecule executing a large amplitude bending vibration. *Molecular Physics*, 27(5):1217–1234, 1974.

[3] Ch. Jungen and A. J. Merer. The Renner–Teller effect. In N. R. Rao, editor, *Molecular Spectroscopy: Modern Research, Volume II*, page 127, 1976.

[4] R. Renner. Zur Theorie der Wechselwirkung zwischen Elektronen-und Kernbewegung bei dreiatomigen, stabförmigen Molekülen. *Zeitschrift für Physik A Hadrons and Nuclei*, 92(3):172–193, 1934.

[5] R. Renner. On the theory of the interaction between electronic and nuclear motion for three-atomic, bar-shaped molecules. In H. Hettema, editor, *Quantum Chemistry: Classical Scientific Papers*, pages 61–80, 2000.

[6] K. Dressler and D. A. Ramsay. The electronic absorption spectra of NH_2 and ND_2. *Philosophical Transactions of the Royal Society of London. Series A, Mathematical and Physical Sciences*, 251:553–602, 1959.

[7] R. N. Dixon. The Renner effect in nearly linear molecule, with application to NH_2. *Molecular Physics*, 9(4):357–366, 1965.

[8] J. M. Brown. The effective Hamiltonian for the Renner–Teller effect. *Journal of Molecular Spectroscopy*, 68(3):412–422, 1977.

[9] J. T. Hougen, P. R. Bunker, and J. W. C. Johns. The vibration-rotation problem in triatomic molecules allowing for a large-amplitude bending vibration. *Journal of Molecular Spectroscopy*, 34:136–172, 1970.

[10] J. W. Cooley. An improved eigenvalue corrector formula for solving the Schrödinger equation for central fields. *Mathematics of Computation*, 15:363–374, 1961.

[11] B. Noumeroff. Méthode nouvelle pour la détermination des orbites et le calcul des éphémérides en tenant compte des perturbations. *Publ. Obs. Cent. Astrophys. Russ.*, 4:1–12, 1923.

[12] B. V. Noumerov. A method of extrapolation of perturbations. *Monthly Notices of the Royal Astronomical Society*, 84(8):592–602, 1924.

[13] J. C. F. Sturm. Mémoire sur la résolution des équations numériques. *Bulletin des Sciences de Férussac*, 11:419–425, 1829.

[14] W. Givens. A method of computing eigenvalues and eigenvectors suggested by classical results on symmetric matrices. *National Bureau of Standards Applied Mathematics Series*, 29:117–122, 1953.

[15] G. Duxbury and R. N. Dixon. The Renner effect in a bent triatomic molecule using the adiabatic approach. *Molecular Physics*, 43(2):255–274, 1981.

[16] G. Duxbury, B. D. McDonald, M. Van Gogh, A. Alijah, Ch. Jungen, and H. Palivan. The effects of vibrational resonances on Renner–Teller coupling in triatomic molecules: The stretch-bender approach. *The Journal of Chemical Physics*, 108(6):2336–2350, 1998.

[17] J. T. Hougen. Rotational energy levels of a linear triatomic molecule in a $^2\Pi$ electronic state. *The Journal of Chemical Physics*, 36(2):519–534, 1962.

[18] A. Alijah and G. Duxbury. Renner–Teller and spin-orbit interactions between the $\tilde{A}^2A_1$ and the $\tilde{X}^2B_1$ states of NH_2: The stretch-bender approach. *Journal of Molecular Spectroscopy*, 211(1):7–15, 2002.

[19] A. Alijah and G. Duxbury. Stretch-bender calculations of the rovibronic energies in the $\tilde{X}^2B_1$ electronic ground state of NH_2. *Journal of Molecular Spectroscopy*, 211(1):16–30, 2002.

[20] G. Duxbury and A. Alijah. Stretch-bender calculations of the rovibronic energies in the excited, $\tilde{A}^2A_1$, electronic state of NH_2 and of the near-resonant high-lying levels of the $\tilde{X}^2B_1$ state. *Journal of Molecular Spectroscopy*, 211(1):31–57, 2002.

[21] P. Jensen, M. Brumm, W. P. Kraemer, and P. R. Bunker. A treatment of the Renner effect using the MORBID Hamiltonian. *Journal of Molecular Spectroscopy*, 171(1):31–57, 1995.

[22] P. Jensen. A new Morse oscillator-rigid bender internal dynamics (MORBID) Hamiltonian for triatomic molecules. *Journal of Molecular Spectroscopy*, 128(2):478–501, 1988.

[23] S. Carter and N. C. Handy. A variational method for the calculation of ro-vibronic levels of any orbitally degenerate (Renner–Teller) triatomic molecule. *Molecular Physics*, 52(6):1367–1391, 1984.

[24] W. H. Green, N. C. Handy, P. J. Knowles, and S. Carter. Theoretical assignment of the visible spectrum of singlet methylene. *The Journal of Chemical Physics*, 94(1):118–132, 1991.

[25] W. Gabriel, G. Chambaud, P. Rosmus, S. Carter, and N. C. Handy. Theoretical study of the Renner–Teller $\tilde{A}^2A_1 - \tilde{X}^2B_1$ system of NH_2. *Molecular Physics*, 81(6):1445–1461, 1994.

[26] W. Gabriel, G. Chambaud, P. Rosmus, S. Carter, and N. C. Handy. Erratum: Theoretical study of the Renner–Teller $\tilde{A}^2A_1 - \tilde{X}^2B_1$ system of NH_2. *Molecular Physics*, 101(12):1933–1933, 2003.

[27] T. Taketsugu and S. Carter. A variational determination of spin-rovibronic energy levels of MgNC in the $A^2\Pi$ state. *Chemical Physics Letters*, 340(3–4):385–389, 2001.

[28] S. Carter, N. C. Handy, P. Rosmus, and G. Chambaud. A variational method for the calculation of spin-rovibronic levels of Renner–Teller triatomic molecules. *Molecular Physics*, 71(3):605–622, 1990.

[29] M. Brommer, B. Weis, B. Follmeg, P. Rosmus, S. Carter, N. C. Handy, H. J. Werner, and P. J. Knowles. Theoretical spin-rovibronic $^2A_1(\Pi_u) - {}^2B_1$ spectrum of the H_2O^+, HDO^+, and D_2O^+ cations. *The Journal of Chemical Physics*, 98(7):5222–5234, APR 1 1993.

[30] S. Carter, N. C. Handy, and R. Tarroni. A variational method for the calculation of spin-rovibronic energy levels of any triatomic molecule in an electronic triplet state. *Molecular Physics*, 103(6–8):1131–1137, 2005.

[31] S. Carter, N. C. Handy, C. Puzzarini, R. Tarroni, and P. Palmieri. A variational method for the calculation of spin-rovibronic energy levels of triatomic molecules with three interacting electronic states. *Molecular Physics*, 98(21):1697–1712, 2000.

[32] R. Tarroni, A. Mitrushenkov, P. Palmieri, and S. Carter. Energy levels of HCN^+ and DCN^+ in the vibronically coupled $X^2\Pi$ and $A^2\Sigma^+$ states. *The Journal of Chemical Physics*, 115(24):11200–11212, 2001.

[33] M. Perić, S. Jerosimić, R. Ranković, M. Krmar, and J. Radić-Perić. An ab initio model for handling the Renner–Teller effect in tetra-atomic molecules. I. Introduction of coordinates and the Hamiltonian. *Chemical Physics*, 330(1–2):60–72, 2006.

[34] M. Perić. A model for the Renner–Teller effect in any linear molecule. *Molecular Physics*, 105(1):59–69, 2007.

[35] M. Perić, M. Petković, and S. Jerosimić. Renner–Teller effect in five-atomic molecules: Ab initio investigation of the spectrum of C_5^-. *Chemical Physics*, 343(2-3):141–157, 2008. Theoretical Spectroscopy and its Impact on Experiment (in honour of Sigrid D. Peyerimhoff).

[36] M. Mitić, R. Ranković, M. Milovanović, S. Jerosimić, and M. Perić. Underlying theory of a model for the Renner–Teller effect in any-atomic linear molecules on example of the $X^2\Pi_u$ electronic state of C_5^-. *Chemical Physics*, 464:55–68, 2016.

CHAPTER 4

First Row Dihydrides

CONTENTS

4.1 INTRODUCTION

From 1970 onwards, the methods of measuring the spectra of free radicals and molecular ions began to separate. One family of experiments was still based upon the detection of absorption and emission spectra, whereas the other group was based upon the use of synchrotron and related instrumentation. Methods of molecular cooling, due the use of supersonic jets, were devised, and many ways of using lasers were developed.

One of the ways of following the development in the use of lasers is to study the application of instrumental developments, in order to improve our understanding of the behaviour of the CH_2 free radical. The first laser-induced fluorescence spectrum of singlet methylene was recorded in 1978 by Welge's group at Bielefeld [1]. They used the output of a Transversely Excited Atmospheric (TEA) CO_2 laser to photo-dissociate acetic anhydride. The dissociation product in the $\tilde{a}\,^1A_1$ state was probed by a dye laser. The fluorescence was then detected by a photomultiplier combined with a gated fast pulse counting system $5\mu s$ wide. Both excitation spectra and laser-induced fluorescence (LIF) spectra were recorded, and identified using the data given by Herzberg and Johns [2].

The next series of measurements, based upon improved experimental techniques, were carried out in the National Research Coucil (NRC) in Ottawa by McKellar, and in the National Institute of Standards and Technology (NIST) Boulder laboratories by the Evenson group [3]. They were followed by measurements made in the University of California, Berkeley, in the Bradley Moore group [4, 5], and in the University of Pennsylvania by the group of Hai-Lung Dai [6, 7].

The Ottawa and NIST-Boulder experiments were based upon "Far Infrared laser magnetic resonance" (LMR) of singlet methylene. This allowed singlet-triplet perturbations, singlet-triplet transitions, and singlet-triplet splitting to be observed. The singlet-singlet and singlet-triplet transitions give rise to LMR spectra because of perturbations between $\tilde{a}\,^1A_1$ and $\tilde{X}\,^1B_1$ states. These effects are shown in Figures 4.1 and 4.2.

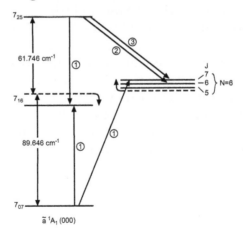

Figure 4.1: Illustration of the energy levels involved in the $7_{25} \rightarrow 7_{16}$ and $7_{16} \rightarrow 7_{10}$ transitions (not drawn to scale). The perturbing triplet level with $N = 6$ is drawn on the right. The dashed lines indicate the positions of the levels before perturbation. The numbers in circles give the number of different laser lines used to observe each transition. A more detailed view of the centre part of this figure is given in Figure 4.2. Reproduced after McKellar et al., *J. Chem. Phys.*, 79:5251, 1983 [3], with the permission of AIP Publishing.

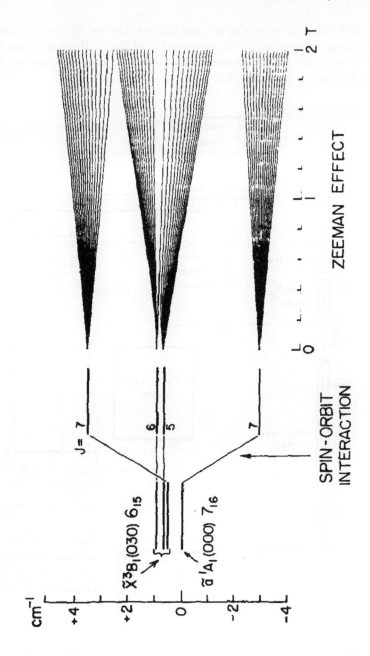

Figure 4.2: Illustration of the interaction between the $\tilde{a}^1 A_1(0,0,0)7_{16}$ level of CH_2 and the perturbing triplet state level with $N = 6$. The nearly linear Zeeman effect of the two $J = 7$ levels at low fields is evident here. Reproduced from McKellar et al., *J. Chem. Phys.* 79:5251, 1983 [3], with the permission of AIP Publishing.

In 1983 Petek, Nesbitt, Ogilby, and Bradley Moore [4] used "Infrared Flash Kinetic spectroscopy" to obtain the spectra of the $^1CH_2$ symmetric (ν_1) and the antisymmetric (ν_3) stretching vibration. A schematic diagram of the instrument is shown in Figure 4.3.

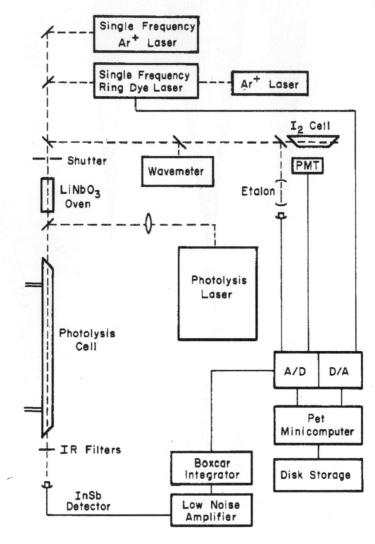

Figure 4.3: Schematic diagram of the high-resolution flash-kinetic spectrometer for measuring absorption and magnetic rotation spectra. The magnet and crossed polarisers are used only for the magnetic rotation experiments. Reprinted with permission from Petek et al., *J. Phys. Chem.* 87:5367, 1983 [4]. Copyright (1983) American Chemical Society.

An example of the use of the flash-kinetic spectrometer for studying absorption spectra is shown in Figure 4.4.

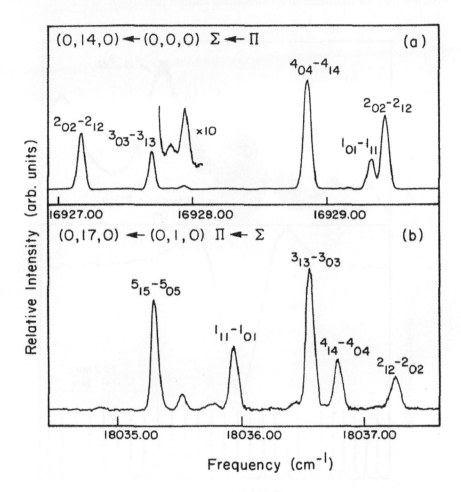

Figure 4.4: Parts of two Q branch spectra recorded by the flash-kinetic spectrometer under typical experimental conditions. (a) Fifteen excimer laser shots are averaged per 19 MHz frequency scan interval. With $\sim$0.2 Torr of CH_2CO and $\sim$0.5 Torr He, there is 30% single pass absorption on the strongest line. A small segment of the spectrum is magnified 10× to indicate the signal to noise level that is typical for most of the transitions observed. (b) The signal on this hot band transition is weaker because the estimated $(0, 1, 0)$ population is only 28% of the total and because the Franck-Condon factor is smaller. One manifestation of the perturbation that affect many $K_a \neq 0\,^1B_1$ states in the triple reversal in the direction of this branch. Reproduced from Petek et al., *J. Chem. Phys.* 86:1172, 1987 [8], with the permission of AIP Publishing.

The next example of the use of the flash-kinetic spectrometer is the study of magnetic rotation spectrum. In the spectrum shown in Figure 4.5, both Zeeman and Zeeman-inactive transitions can be identified.

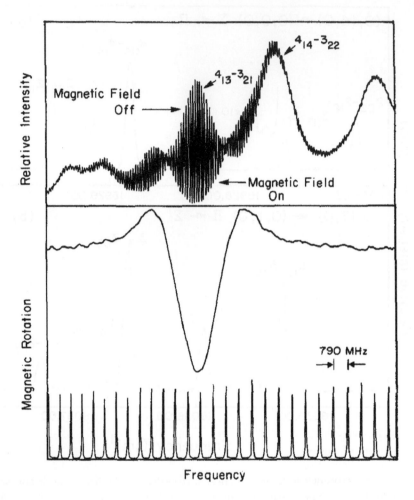

Figure 4.5: Magnetic-rotation spectrum taken with 22 kHz scan step size shows a strongly Zeeman active transition and several inactive transitions. The fast square-wave oscillation is produced by modulation of the magnetic field between 0 and 1.1 kGauss. When the magnetic field is off, only the transient absorption is observed; when the magnetic field is on, the transient absorption and magnetic rotation signals are superimposed. The modulation amplitude plotted in the lower frame gives the deconvoluted magnetic rotation lineshape of the $(0, 13, 0) - (0, 0, 0)$ $4_{13} - 3_{21}$ transition. The Zeeman effect is in the 4_{13} level. Reproduced from Petek et al., *J. Chem. Phys.* 86:1172, 1987 [8], with the permission of AIP Publishing.

The characteristic line shape of the Zeeman-active transition results from a phaseshift between the two circular components of linearly polarized light, as analysed in Figure 4.6.

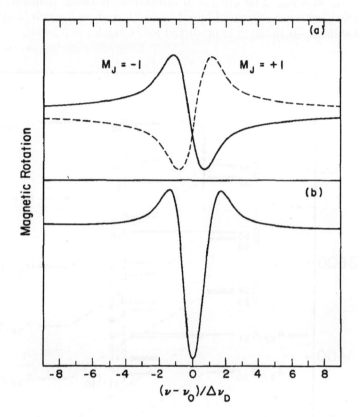

Figure 4.6: The calculated magnitude and line shape of the magnetic-rotation signals observed in this study result from a phaseshift between the two circular components of linearly polarized light due to an imbalance between real parts of the index of refraction for $\Delta M_J = +1$ and $\Delta M_J = -1$ transitions. For simplicity, consider an R-branch transition from a triplet-perturbed $J = 1$ lower level to an unperturbed $J = 2$ upper level. (a) Dispersion curves for $M_J = +1$, $\Delta M_J = \pm 1$ (—) and $M_J = -1$, $\Delta M_J = \pm 1$ (— — —). There is no contribution for $\Delta M_J = 0$, since the $\Delta M_J = \pm 1$ transitions have the same intensity and, thus, the net rotation cancels exactly. In the presence of a magnetic field, the lower-state $\Delta M_J = \pm 1$ levels and the corresponding dispersion curves are tuned apart. (b) The net rotation from $M_J = \pm 1$, $\Delta M_J = \pm 1$ transitions is the sum of the two curves in (a) and is in qualitatively excellent agreement with experimental observation (see Figure 4.5). For the purpose of this calculation, the ratio of homogeneous to Doppler linewidth is 0.06, characteristic of a 400 K distribution with 50 MHz pressure broadening. Reproduced from Petek et al., *J. Chem. Phys.* 86:1172, 1987 [8], with the permission of AIP Publishing.

The next examples that are studied are singlet-triplet interactions with different selection rules, $\Delta K_a = 0$ and $\Delta K_a = \pm 2$; see Figure 4.7. The $\Delta K_a = 2$ interaction is only important for near-resonant interactions, such as indicated for 4_{31}, 4_{32}, and 6_{33}. The $\Delta K_a = 0$ interaction is much stronger, however, the coupled levels are separated by ~ 140 cm^{-1}, and, as a result, a systematic upward shift of ~ 0.15 cm^{-1} is predicted for $^1CH_2$ $K_a = 3$ levels.

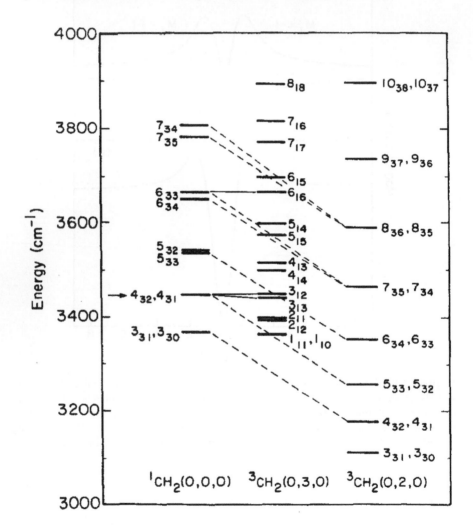

Figure 4.7: The interactions of $^1A_1$ $(0,0,0)$ $K_a = 3$ levels with $^3B_1$ $(0,3,0)$ $K_a = 1$ and $(0,2,0)$ $K_a = 3$. Reproduced from Petek et al., *J. Chem. Phys.* 86:1189, 1987 [5], with the permission of AIP Publishing.

Another type of interaction is shown in Figure 4.8.

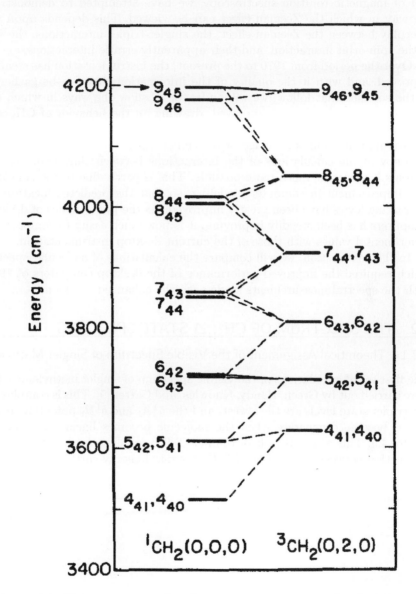

Figure 4.8: The interactions of $^1A_1\,(0,0,0)\,K_a = 4$ levels with $^3B_1\,(0,2,0)$. Reproduced from Petek et al., *J. Chem. Phys.* 86:1189, 1987 [5], with the permission of AIP Publishing.

In this summary of the behavior of far infrared laser magnetic resonance, and of magnetic rotation spectroscopy, we have attempted to demonstrate the way in which the Zeeman effect can be viewed. This depends upon the interplay between the Zeeman effect, the singlet-triplet interactions, the size of the spin-orbit interaction, and their apparently erratic interactions.

Over the period from 1970 to the present, the instrumentation has steadily improved, and with it the quality of the information that can be gathered. In the sections that follow, we will endeavor to show the ways in which our knowledge of molecules has improved, drawing on the behavior of CH_2 and NH_2 as examples.

Not only has our knowledge of instrumentation improved, but also the accuracy of the calculations of the interactions between singlet and triplet systems has become much more possible. This is partly due to the considerable improvement in computer modeling, so that the predicted locations of interacting levels have been greatly improved. As the performance of desktop computers has been steadily improving, it is now much easier to compare the experimental values with those of the current desktop instrumentation.

In the next section, we will compare the calculations of molecular spectra, which required the improved performance of the desktop computers of 1991, with the spectral measurements made with the enhanced spectrometers.

4.2   THE SPECTRUM OF CH_2, Δ STATE SPLITTING

4.2.1   Theoretical Assignment of the Visible Spectrum of Singlet Methylene

The first set of calculations of the visible spectrum of singlet methylene, CH_2, were carried out by Green, Handy, Knowles, and Carter [9]. This is complex, as the triplet state levels are the lowest, and the $\tilde{a}^1 A_1$ and $\tilde{b}^1 B_1$ potential energy curves become degenerate when the molecule becomes linear, as shown in Figure 4.9. This is an example of Renner–Teller coupling, leading to a strong interaction between the $\tilde{a}^1 A_1$ and $\tilde{b}^1 B_1$ potential energy surfaces.

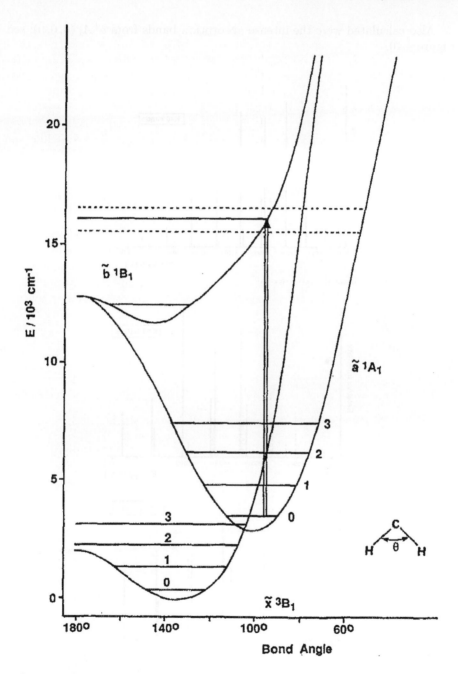

Figure 4.9: Theoretical assignment of the visible spectrum of singlet methylene. The lowest-lying potential energy surfaces of CH_2, plotted vs. the bend angle. The two lowest-lying singlet states become degenerate at linearity, and interact strongly through Renner–Teller coupling. Reproduced from Green et al., *J. Chem. Phys.* 94:118, 1991 [9], with the permission of AIP Publishing.

Also calculated were the intense absorption bands from $\tilde{a}^1 A_1 (0,0,0)$; see Figure 4.10.

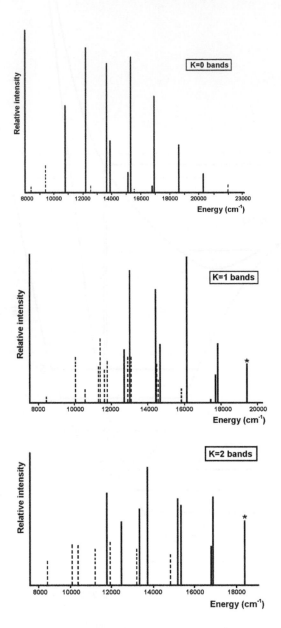

Figure 4.10: Predicted intense absorption bands from, $\tilde{a}^1 A_1 (0,0,0)$ (a) $K = 0$, (b) $K = 1$ and (c) $K = 2$. Bands marked * are predicted to be split by near resonances. Reproduced from Green et al., *J. Chem. Phys.* 94:118, 1991 [9], with the permission of AIP Publishing.

The final example of problems studied by Green and his collaborators was the behavior of the vibrational wavefunctions, as a function of the bend angle and of the symmetric stretch. The resultant calculations of the behavior are plotted in Figure 4.11.

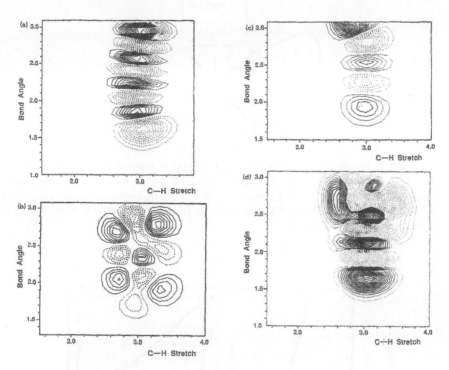

Figure 4.11: Vibrational wave function vs. the symmetric stretch and the bend angle. This plot shows 4 of the 6 contour plots of Green Jr. et al. (a) Regular nodal pattern, (b) identifiable, (c) identifiable, and (d) nodal patterns of contour plot are so perturbed that it is difficult to identify the vibrational quantum numbers. Reproduced from Green et al., *J. Chem. Phys.* 94:118, 1991 [9], with the permission of AIP Publishing.

4.2.2 Determination of Radiative Lifetimes of CH_2 $(\tilde{b}\,^1B_1)$

The improved technical capability of the instrumentation has opened new ways in which the radiative lifetimes of the excited singlet state of CH_2, $\tilde{b}\,^1B_1$, could be explored. The first example is the work of I. García-Moreno, E. R. Lovejoy, C. Bradley Moore, and G. Duxbury [10], which is presented in Figures 4.12 to 4.14.

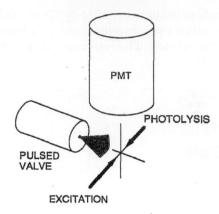

Figure 4.12: Experimental excitation and detection geometry. The distance between the window of the Photo-Multiplier Tube (PMT) (16 cm^2 active area) and the interaction zone was 5 cm. Excitation was 3.5 cm downstream from the pulsed valve. Reproduced from García-Moreno et al., *J. Chem. Phys.* 98:873, 1993 [10], with the permission of AIP Publishing.

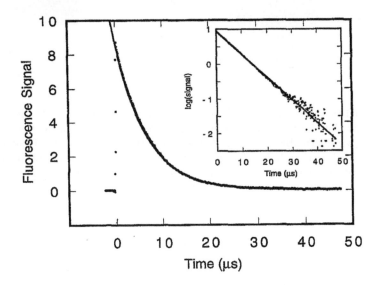

Figure 4.13: The temporal evolution of the fluorescence from CH_2 following the excitation of the $(\tilde{b}^1 B_1)\,(0, 15, 0)\,1_{11} \leftarrow (\tilde{a}^1 A_1)\,(0, 0, 0)\,2_{21}$ transition. The solid line is the fit to the single exponential decay, giving a lifetime of 5.8 μs. The inset shows the logarithmic representation of the data and fit. Reproduced from García-Moreno et al., *J. Chem. Phys.* 98:873, 1993 [10], with the permission of AIP Publishing.

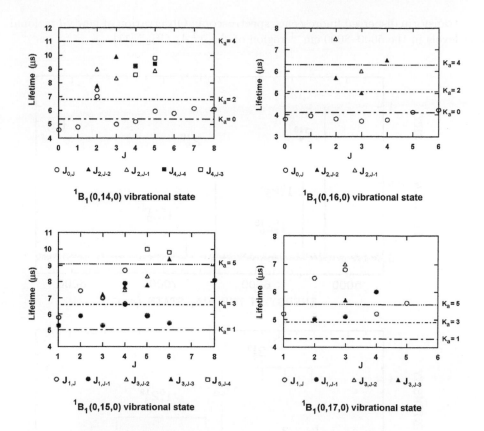

Figure 4.14: Radiative lifetimes as a function of J. The dashed lines represent the calculated vibronic lifetimes of $K_a = 0$, 2, and 4 for $(0, 14, 0)$ and $(0, 16, 0)$, and $K = 1$, 3, and, 5 for $(0, 15, 0)$ and $(0, 17, 0)$. The variation of the observed rovibronic lifetimes is due to the lifetime lengthening. In part, it is owing to Renner–Teller coupling of the $\tilde{b}\,^1B_1$ to the $\tilde{a}\,^1A_1$ state, and also singlet-coupling, producing a marked and erratic variation of the rotational state. Reproduced from García-Moreno et al., *J. Chem. Phys.* 98:873, 1993 [10], with the permission of AIP Publishing.

4.2.3 The Group of Hai-Lung Dai, Department of Chemistry, The University of Pennsylvania

The group of Hai-Lung Dai, in the Department of Chemistry at The University of Pennsylvania, Philadelphia, Pennsylvania, used different experimental methods from those developed in C. Bradley Moore's group at Berkeley. The first paper of Hartland, Wei Xie, Dong Qin, and Hai-Lung Dai, in 1992, was devoted to the study of "Strong asymmetry induced $\Delta K_a = 3$ transitions in the CH_2 $(\tilde{b}\,^1B_1 - \tilde{a}\,^1A_1)$ spectrum: a study by Fourier transform emission spectroscopy" [6]. Their second paper in 1993, Figure 4.15, concerned "Fourier transform dispersed fluorescence spectroscopy: Observation of new vibrational levels in the 5000–8000 cm^{-1} region of the $\tilde{a}\,^1A_1$ CH$_2$" [11].

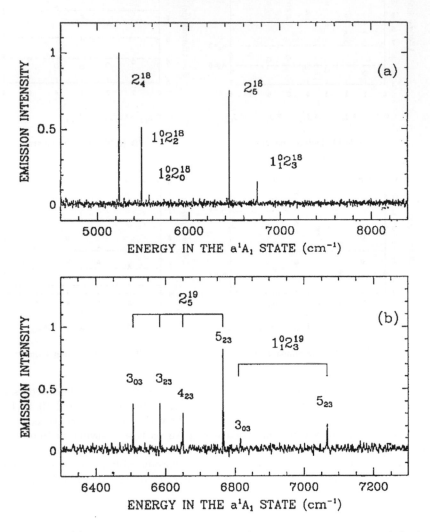

Figure 4.15: (a) Fourier transform dispersed fluorescence spectra recorded after excitation of the CH$_2$ $\tilde{b}^1B_1$ $(0, 18^0, 0)$ $J_{K_aK_c} = 0_{00}$ level. Five $\tilde{b} \rightarrow \tilde{a}$ vibrational bands can be seen and are labeled in the spectra. Each band consists of a single rotational transition $0_{00} \rightarrow 1_{10}$. (b) Fourier transform dispersed fluorescence spectra recorded after excitation of the CH$_2$ $\tilde{b}^1B_1$ $(0, 19^1, 0)$ $J_{K_aK_c} = 4_{13}$ level. Two $\tilde{b} \rightarrow \tilde{a}$ bands can be seen: 2_5^{19} and $1_1^0 2_3^{19}$. The rotational levels in the $\tilde{a}^1A_1$ state are labeled in the figure. Reproduced from Hartland et al., *J. Chem. Phys.* 98:2469, 1993 [11], with the permission of AIP Publishing.

Their final paper, published in 1995, is a study of the "Renner–Teller effect on the highly excited bending levels of $\tilde{a}^1A_1$ CH$_2$" [7].

4.2.4 Groups of Trevor J. Sears and Gregory E. Hall at Brookhaven National Laboratory, Long Island

In the period since 1994, much of the information about the spectrum of CH$_2$ has originated in the groups of Trevor J. Sears and Gregory E. Hall at Brookhaven National Laboratory, Long Island. These experiments have switched to the infrared region from the visible region of Feldman et al. [1], Petek and his colleagues [4, 5, 8], Green, Handy, Knowles, and Carter [9], García-Moreno and her collaborators [10, 12], and Hai-Lung Dai, G. V. Hartland, and their team [6, 7, 11].

One of the first examples of this was the analysis of the "Near-infrared vibronic spectrum of the CH$_2$ $\tilde{b}^1B_1 - \tilde{a}^1A_1$ transition" [13]. In order to improve the efficiency in the near-infrared region, Sears and Hall and their collaborators constructed a Ti:sapphire, ring laser-based, transient absorption spectrometer, with Doppler limited resolution. They used this spectrometer to measure part of the $\tilde{b} - \tilde{a}$ electronic spectrum, which lies between 11400 cm^{-1} and 12500 cm^{-1}.

The CH$_2$ radicals were formed by 308 nm photolysis of ketene (CH$_2$CO) using an excimer laser. They identified 8 vibronic bands originating in the vibrationless level of the $\tilde{a}^1A_1$ state. Their results were similar to those of the Bradley Moore group. Bloch, Field, Hall, and Sears [14] also demonstrated a frequency modulation (FM) technique that improved the sensitivity of their transient absorption spectrometer [15].

The singlet ($\tilde{a}^1A_1$ state) CH$_2$ radicals were produced by photolysis of ketene in a slow flow cell. The transient absorption spectrum was obtained by scanning a tunable near-infrared (near-IR) spectrometer between 11400 cm^{-1} and 12500 cm^{-1}. More recently, the frequency modulation technique has been used to provide higher sensitivity for detecting weak signals. The experimental setup is shown in Figure 4.16.

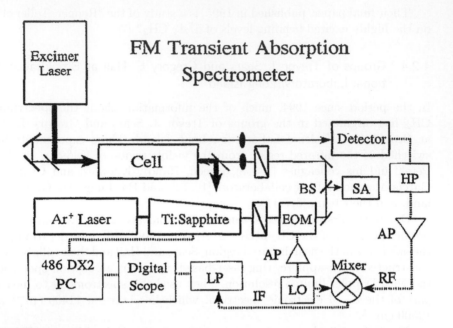

Figure 4.16: Block diagram of the frequency modulation (FM) experimental setup. EOM = electro-optic modulator, RF = radio frequency, IF = intermediate frequency. LO = local oscillator, LP = low-pass filter, HP = high-pass filter, BS = beam splitter, SA = spectrum analyser, and AP = amplifier. Reproduced from Chang et al., *J. Chem. Phys.* 101:9236, 1994 [13], with the permission of AIP Publishing.

In these experiments, Chang et al. measured the near IR spectrum of the CH$_2$ $\tilde{b}^1 B_1 - \tilde{a}^1 A_1$ transition between 11400 cm^{-1} and 12500 cm^{-1}. They identified 163 rovibronic transitions in 8 bands. These assignments were based upon the rotational combination differences in the $\tilde{a}(0,0,0)$ level. Where an overlap with the data of Herzberg and Johns exists, the results are consistent, but the new data have higher precision.

Two vibronic bands were observed for the first time. The upper levels of these new vibronic bands possess Π vibronic symmetry, and are located about 2000 cm^{-1} above the barrier to linearity. The use of FM modulation improved the signal to noise ratio by a factor of at least 20 above that observed in Figure 4.17.

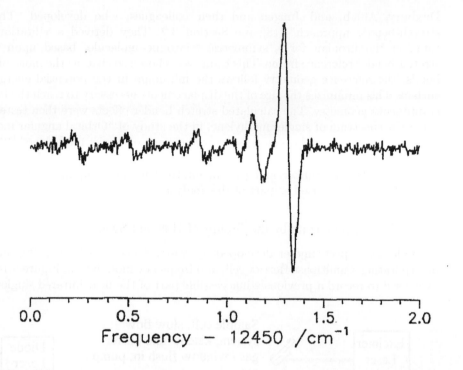

Figure 4.17: FM spectrum of the CH$_2$ $\tilde{b} \leftarrow \tilde{a}$ transition with 10 excimer laser shots averaged per 100 MHz scan increment. Reproduced from Chang et al., *J. Chem. Phys.* 101:9236, 1994 [13], with the permission of AIP Publishing.

4.2.5 Improved Computational Methods

In the period from 1998 onwards, several groups extended the methods of calculating the role of vibrational resonances, and Renner–Teller coupling in the behavior of CH$_2$ and NH$_2$. In the calculations on the behavior of CH$_2$, two groups played a major role. One is the group of Gu, Hirsch, Buenker, Brumm, Osman, Bunker, and Jensen [16], referred to in the following as Gu et al. P. R. Bunker and P. Jensen have been the main drivers of the development of methods. They made a theoretical study of the absorption spectrum of singlet CH$_2$, in which Renner effects are included and play a major role. Much of their development is included in the monograph of P. R. Bunker and P. Jensen, *Molecular Symmetry and Spectroscopy*, 2nd Edition, NRC Research Press, Ottawa, Canada [17]. The section on the Renner effect and the Jahn-Teller effect is particularly interesting. The description begins on page 370 with the Renner effect, followed on page 379 with the Jahn-Teller effect. Pages 410 to 413 give some bibliographical notes.

The other group that has made a considerable contribution is that of

Duxbury, Alijah, and Jungen and their colleagues, who developed "The stretch-bender approach" [18]; see Section 3.2. They derived a vibration-rotation Hamiltonian for a symmetric triatomic molecule, based upon a stretch-bender reference frame. This frame was chosen so that, as the molecule bends, the reference geometry follows the minimum in the potential energy surface. This minimizes the size of the displacements necessary to reach the instantaneous geometry. The calculated stretch bender effects were then tested using the spectrum of singlet methylene, in the study of "Orbital angular momentum and vibrational resonances in the spectrum of singlet methylene" [19].

In the following experiments on CH_2 carried out by the group of Hall and Sears, both the Gu, Bunker, and Jensen, and the Duxbury, Alijah, and Jungen methods, have been used as part of the analysis.

4.2.6 New Experiments by the Groups of Hall and Sears

A diode laser spectrometer developed by Marr, Sears, and Chang [20] and incorporating a multipass Herriot cell, and frequency modulation, Figure 4.18, was used to record a previously inaccessible part of the near infrared singlet-

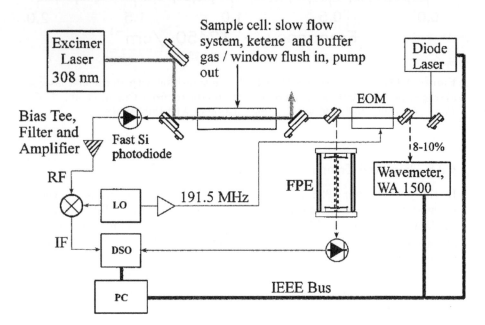

Figure 4.18: Block diagram of the experimental setup. EOM = electro-optic modulator, FPE = Fabry-Perot etalon, LO = local oscillator, RF = radio frequency, IF = intermediate frequency, DSO = digital storage oscilloscope, and PC = personal computer. Reproduced from Marr et al., *J. Chem. Phys.* 109:3431, 1998 [20], with the permission of AIP Publishing.

singlet absorption spectrum of methylene between 10000 and 10600 cm^{-1}; see Figures 4.19 and 4.20.

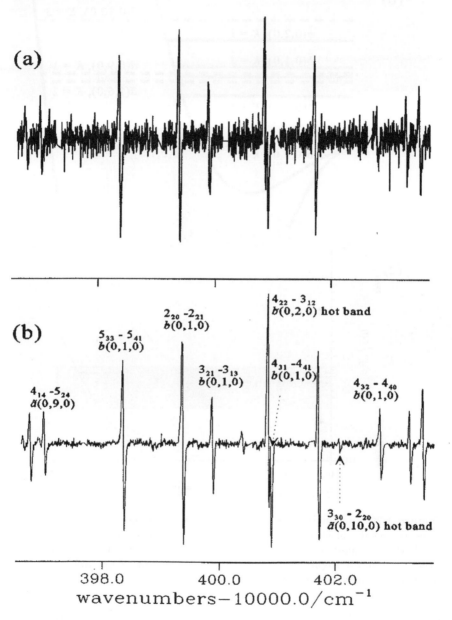

Figure 4.19: Section of raw CH_2 spectrum recorded in the region of 10400 cm^{-1}. (a) FM only, and (b) FM with multipass Herriott cell, identified transitions are labelled. Reproduced from Marr et al., *J. Chem. Phys.* 109:3431, 1998 [20], with the permission of AIP Publishing.

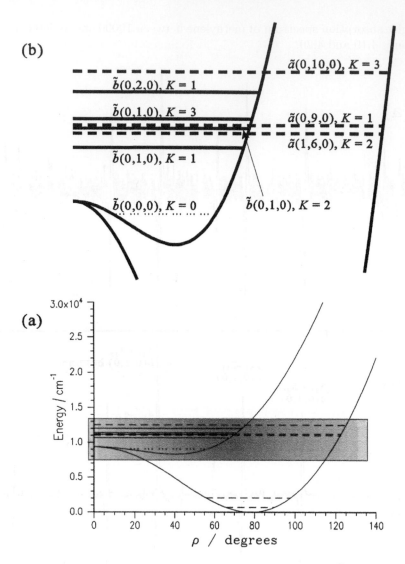

Figure 4.20: (a) Bending potential energy curves for the $\tilde{a}$ and $\tilde{b}$ states computed from the results of Duxbury and Jungen [21]. (b) Boxed section of (a), expanded to show states most recently identified in this, and the previous near-IR diode laser study, was shown. Reproduced from Marr et al., *J. Chem. Phys.* 109:3431, 1998 [20], with the permission of AIP Publishing.

The next experiments, in which longer wavelengths near 9500 cm^{-1} were surveyed, were carried out by Kobayashi, Pride, and Sears [22]. They made a cut through the potential energy surfaces for the $\tilde{a}$ and $\tilde{b}$ states of CH_2 along the bending angle coordinate. These were computed from the results of Duxbury and Jungen [21]. The lowest excited state level that was then identified was the $\tilde{b}\,(0,0,0)\,K=0$. The Renner–Teller effect and the spin-orbit coupling between the ground triplet and the singlet states have a considerable effect. In addition, local an-harmonic and Coriolis interactions can be large. A combination of the interactions of the two types of coupling severely perturb both the $\tilde{a}$ and the $\tilde{b}$ state ro-vibrational energy levels, causing them to exhibit very irregular spacings; see Figure 4.21.

(a)

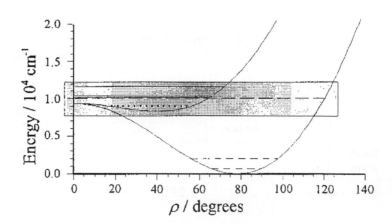

(b)

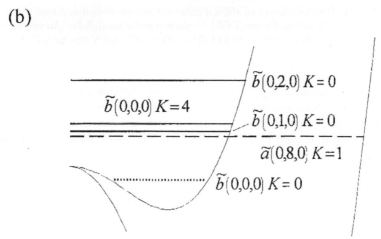

Figure 4.21: (a) A cut through the potential energy surfaces for the $\tilde{a}$ and $\tilde{b}$ states of CH_2, along the bending angle coordinate, computed from the results of Duxbury and Jungen. (b) Boxed section of (a) expanded to show the selected $\tilde{b}$ levels clearly. Reprinted with permission from Kobayashi et al., *J. Phys. Chem. A* 104:10119, 2000 [22]. Copyright (2000) American Chemical Society.

The next important step was the measurement by "Sub-Doppler Spectroscopy" of mixed state levels in CH_2, by Chih-Hsuan Chang, G. E. Hall, and T. J. Sears [23]. Figure 4.22 shows the experimental setup.

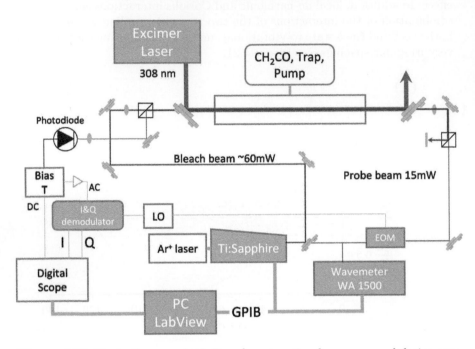

Figure 4.22: Block diagram of sub-Doppler saturation frequency-modulation spectrometer. LO = local oscillator. EOM = electro-optic modulator. Reproduced from Chang et al., *J. Chem. Phys.* 133:144310, 2010 [23], with the permission of AIP Publishing.

Perturbations in the 7_{16} and 8_{12} mixed singlet/triplet levels of $\tilde{a}^1 A_1$ (0,0,0) methylene have been reinvestigated by frequency-modulated laser sub-Doppler saturation spectroscopy. The hyperfine structure was completely resolved for both the predominantly singlet and the predominantly triplet components of these mixed rotational levels, Figures 4.23 to 4.25.

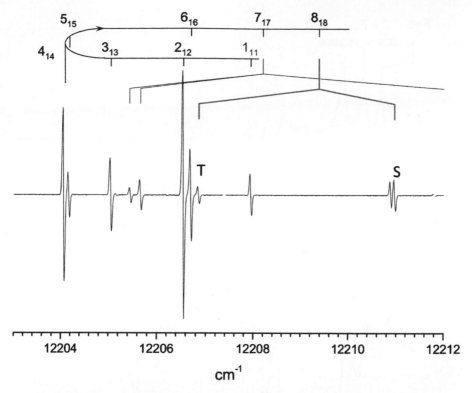

Figure 4.23: Section of the $^PQ_1$ branch of the $\tilde{b}^1 B_1\,(0,3,0)-\tilde{a}^1 A_1\,(0,0,0)$ transition of the CH_2 radical in the region of 12206 cm^{-1} with the rotational assignments. The symbols S and T indicate the "singlet" and "triplet" components of the perturbed 8_{18} rotational level. Reproduced from Chang et al., *J. Chem. Phys.* 133:144310, 2010 [23], with the permission of AIP Publishing.

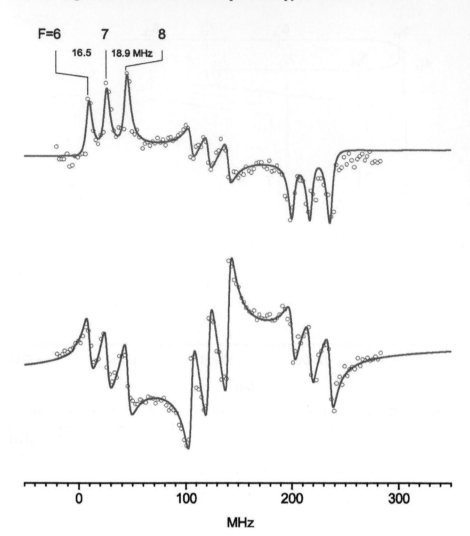

Figure 4.24: Hyperfine-resolved absorption and dispersion spectra of the "singlet" component of the perturbed 7_{16} rotational level. Reproduced from Chang et al., *J. Chem. Phys.* 133:144310, 2010 [23], with the permission of AIP Publishing.

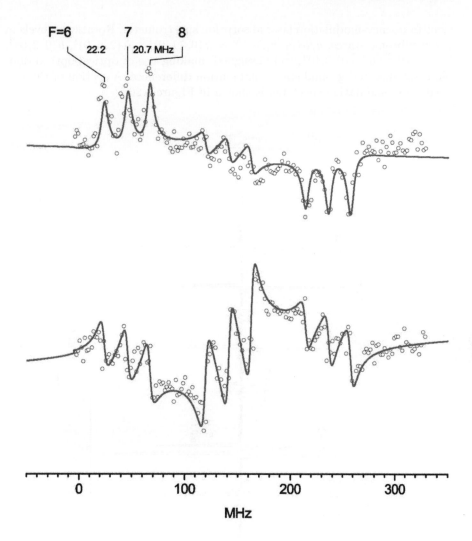

Figure 4.25: As in Figure 4.24, but for the "triplet" component of the perturbed 7_{16} rotational level. The nearly equal hyperfine splitting in the "singlet" and "triplet" components indicates almost complete mixing. Reproduced from Chang et al., *J. Chem. Phys.* 133:144310, 2010 [23], with the permission of AIP Publishing.

While much of the singlet absorption spectrum of methylene, CH_2, has now been recorded using high resolution laser-based methods, there are still some holes in the coverage between the visible and near infrared regions. In this paper [24], (*J. Mol. Spectrosc.*, 267:50-57, 2011), C.-H. Chang, Z. Wang, G. E. Hall, T. J. Sears, and Ju Xin describe transient laser absorption spectroscopy experiments. Bands in the CH_2 $\tilde{b}^1B_1 - \tilde{a}^1A_1$ transition between 12500 cm^{-1} and 13000 cm^{-1} were recorded at Doppler limited resolution using a transient frequency-modulation laser absorption spectrometer. Rotational levels in seven vibronic states, $\tilde{a}/\tilde{b}\,(v_1, v_2, v_3)^K = \tilde{a}\,(0, 11, 0)^{1,3}$, $\tilde{a}\,(2, 6, 0)^1$, $\tilde{b}\,(0, 2, 0)^3$, $\tilde{b}\,(1, 1, 0)^{1,2}$ and $\tilde{b}\,(0, 3, 0)^1$, were assigned, making use of optical-optical double resonance and ground state combination differences. A section of the raw frequency modulation spectrum is shown in Figure 4.26.

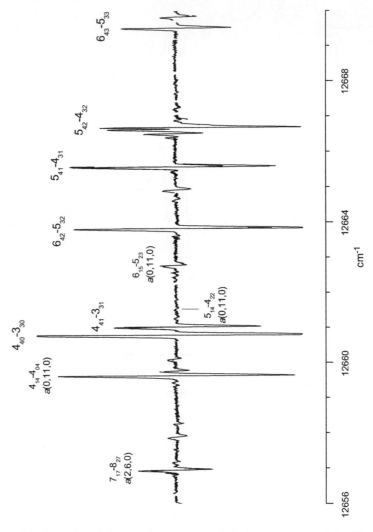

Figure 4.26: A section of the raw frequency modulation spectrum of the $^rR_{3J-2}$ and $^rR_{3J-3}$ branches of the CH$_2$ $\tilde{b}\,(0, 2, 0)^{K=4} - \tilde{a}\,(0, 0, 0)$ transition in the 12660 cm^{-1} region, with rotational assignments. Reprinted from Chang et al., *J. Mol. Spectrosc.* 267:50, 2011 [24]. Copyright (2011), with permission from Elsevier.

Rotational term values of newly assigned vibrational levels are plotted in Figure 4.27 as a function of $J(J+1)$, while in Figure 4.28 a comparison is made between the observed and calculated [16] rotational energy levels.

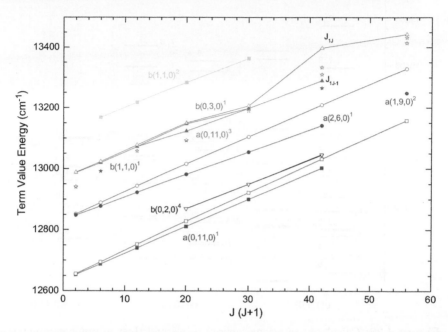

Figure 4.27: Rotational term values of the assigned vibrational levels in the region of 12500-13000 cm^{-1}. Open and closed symbols depict the K-doublets. Reprinted from Chang et al., *J. Mol. Spectrosc.* 267:50, 2011 [24]. Copyright (2011), with permission from Elsevier.

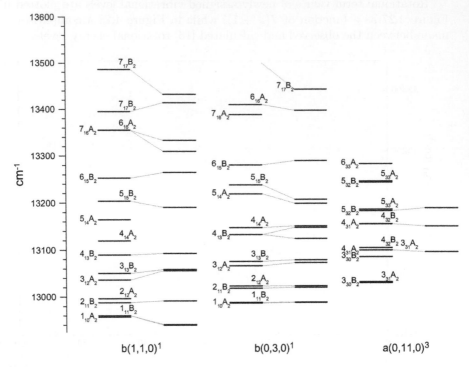

Figure 4.28: Calculated rotational energy levels and their vibronic symmetry for the $\tilde{b}\,(0,1,0)^1$, $\tilde{b}\,(0,3,0)$ and $\tilde{a}\,(0,11,0)$ vibronic levels. Available corresponding experimental values are plotted to the right. Reprinted from Chang et al., *J. Mol. Spectrosc.* 267:50, 2011 [24]. Copyright (2011), with permission from Elsevier.

4.2.7   The CH_2 $\tilde{b}^1B_1 - \tilde{a}^1A_1$ Band Origin at $1.20\,\mu m$

C. H. Chang, J. Xin, T. Latsha, E. Otruba, Z. Wang, G. E. Hall, T. J. Sears, and B.-C. Chang, *J. Phys. Chem.* A, 115:9440-9446, 2011

The spectrum of the $\tilde{b}^1B_1 - \tilde{a}^1A_1$ band system of methylene has been the focus of many experimental and theoretical studies since the initial report by Herzberg and Johns [2]. At high resolution, the rotational structure exhibits many perturbations. These perturbations are associated with the Renner–Teller effect, the occurrence of anharmonic resonances, and to the spin-orbit coupling between the three low lying electronic states, $\tilde{b}^1B_1$, $\tilde{a}^1A_1$, and $\tilde{X}^3B_1$.

Much of the spectrum has now been mapped out by the Moore group [4, 5, 8, 12] in the visible region, and by Sears and Hall and their colleagues using laser absorption spectroscopy [13–15, 20, 22]. Theoretical and computational work by Kalemos, Dunning, Mavridis, and Harrison [25], Gu et al. [16], and by Alijah, Duxbury, and Jungen [18, 19] has provided insights into the cause of spectral complexity, and has guided much of the experimental work.

Despite these efforts, the spectrum continues to defy a complete assignment, owing the dense background of rotational levels belonging to the $\tilde{a}^1A_1$ and $\tilde{X}^3B_1$ vibronic states. These contaminate the pure $\tilde{b} - \tilde{a}$ transitions. In addition, the precision of the computed spectrum is insufficient to resolve ambiguities in assignments, even where the level densities are lower.

The region of combined $\tilde{a}^1A_1$ and $\tilde{b}^1B_1$ state potentials represents the first and simplest occurrence of a periodic resonance phenomenon first noted by Duxbury and Jungen [21]. The ladders of vibronic K_a belonging to the two electronic states come into resonance owing to their different spacings. These resonances lead to level shifts and spectral intensity borrowing that contribute to the observed complexity throughout the spectrum. A section of the raw spectrum of the $^PQ_1(J)$ branch near 8336 cm^{-1} and its assignment is shown in Figure 4.29.

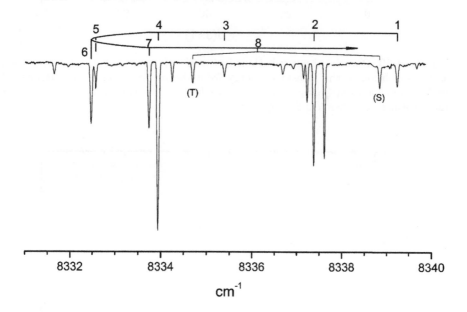

Figure 4.29: Rotational assignment of the $^PQ_1(J)$ branch of the band origin near 8336 cm^{-1}. The (S) and (T) labels signify transitions from the mixed singlet/triplet lower states terminating at the same 8_{08} upper level. Reprinted with permission from Chang et al., *J. Phys. Chem. A* 115:9440, 2011 [26]. Copyright (2011) American Chemical Society.

Upper state rotational energies were fitted to a model Hamiltonian; the molecular parameters are given in Table 4.1. Perturbed levels can then be identified due to the deviations of their observed rotational energies from that of the model Hamiltonian. Two examples are shown in Figures 4.30 and 4.31.

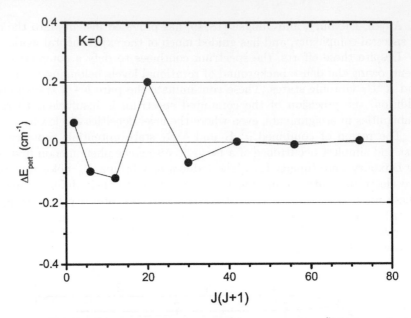

Figure 4.30: Rotational perturbations in the $K = 0$ levels of $\tilde{b}^1 B_1$ $(0, 0, 0)$, shown as the deviation of the measured energy from the pseudolinear model. Reprinted with permission from Chang et al., *J. Phys. Chem. A* 115:9440, 2011 [26]. Copyright (2011) American Chemical Society.

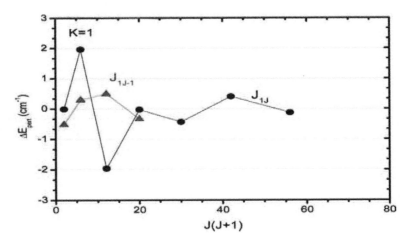

Figure 4.31: Rotational perturbations in the $K = 1$ levels of $\tilde{b}^1 B_1$ $(0, 0, 0)$, shown as the deviation of the measured energy from the pseudolinear model. Reprinted with permission from Chang et al., *J. Phys. Chem. A* 115:9440, 2011 [26]. Copyright (2011) American Chemical Society.

Table 4.1: Molecular Parameters for the Upper Levels $(\text{cm}^{-1})^{a,b}$

| | $\tilde{b}\,(0,0,0)$ | | |
	$K = 0$	$K = 1$	$K = 1^c$
B_v	7.6890(94)	7.537(75)	7.673(37)
$D_v \times 10^2$	0.186(32)	-2.56(33)	-1.98(17)
$H_v \times 10^4$	0.212(30)	-3.67(388)	-3.07(19)
q^d		0.351(22)	0.376(12)
ν_0	8350.971(65)	8376.66(43)	8376.01(20)
σ	0.104	0.816	0.357

[a] One standard deviation limit in parentheses.
[b] $K_a = 0$ and $K_a = 1$ lines were fitted separately.
[c] For this column, the transitions terminating in the 2_{12} and 3_{13} upper levels were not included in the fitting.
[d] Energies of the $K_a = 1$, K-doublet levels are $\pm qJ(J+1)$, with the upper sign for levels with $K_c = J$.

In Table 4.2, the experimentally derived molecular constants are compared with theoretical data. Finally, Figure 4.32 shows the rotational level structure around the barrier to linearity.

Table 4.2: Comparison of the Experimental and Calculated ν_0, B_v and q_v Values (cm^{-1}) of the $\tilde{b}\,(0,0,0)$ States of CH_2

		Chan et al. [26]	Gu et al. [16]	Duxbury et al. [19]	Green et al. [9]
$K = 0$	ν_0	8350.971	8354.034	8363.8	8383
	B_v	7.689		7.67	
$K = 1$	ν_0	8376.01	8400.853	8413.0	8424
	B_v	7.673		8.18	
	q			0.47	

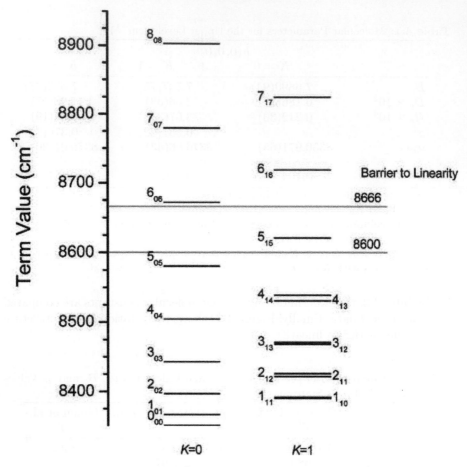

Figure 4.32: Term values (cm^{-1}) of rotational levels in the $K = 0$ and $K = 1$ levels with calculated barriers to linearity, which corresponds to the degenerate $^1\Delta_g$ state at linear geometry, Ref. [7] (8600), Ref. [16] (8666). Reprinted with permission from Chang et al., *J. Phys. Chem. A* 115:9440, 2011 [26]. Copyright (2011) American Chemical Society.

4.3 THE SPECTRUM OF NH$_2$, Π STATE SPLITTING

The other molecule that has been studied in parallel with CH$_2$ is NH$_2$. As was noted in the introduction, a detailed analysis of the Renner–Teller effect in the spectrum of NH$_2$ and ND$_2$ was given by Dressler and Ramsay in 1959 [27]. This work was followed up by Johns, Ramsay and Ross in 1976 in a study of the $\tilde{A}^2A_1 - \tilde{X}^2B_1$ absorption spectrum of NH$_2$ between 6250 and 9500 Å, [28], and by studies of the spectrum of ND$_2$ by Muenchausen et al. [29].

One of the reasons for studying the spectrum of NH$_2$ in some detail is the effects of spin-orbit coupling on the interactions between the $\tilde{A}^2A_1$ and $\tilde{X}^2B_1$

states, in addition to the Renner–Teller coupling. This may involve the use of Fourier transform methods, optogalvanic methods, and new approaches to the separation of differing isotopologues. One of the first examples of Fourier transform spectroscopy was given by Vervloet [30], in his combination of laser excited fluorescence and Fourier transform spectroscopy to study the electronic ground state of NH_2. Following on from this, Dixon, Irving, Nightingale, and Vervloet [31] recorded electronic spectra of NH_2 by three methods: as LIF excitation spectra of the products of photolysis of NH_3 in its $\tilde{A} - \tilde{X}$ bands; as dispersed-emission spectra following laser excitation with this same source; and as emission spectra from a discharge through ammonia recorded with a Fourier-transform spectrometer. In 1999, a study by Hadj Bachir, Huet, Destombes, and Vervloet [32] was published, in which a combined analysis of laser optogalvanic and Fourier transform emission spectra of NH_2 close to the barrier to linearity was presented.

4.3.1   Spectrum obtained by Vervloet and colleagues at Lille University

Figure 4.33 shows a part of their optogalvanic spectrum.

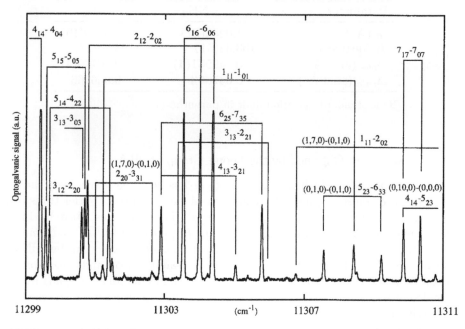

Figure 4.33: A small part of the near-infrared optogalvanic spectrum of NH_2. Lines without vibrational assignment belong to the subband $\Pi\,(0,0,0)\tilde{A}\,^2A_1 - (0,0,0)\tilde{X}\,^2B_1$. Reprinted from Hadj Bachir et al., *J. Mol. Spectrosc.* 193:326, 1999 [32]. Copyright (1999), with permission from Elsevier.

4.3.2 Rotational Spectrum of the NH_2, NHD, and ND_2 Radicals Observed by Morino and Kamaguchi

Morino and Kawaguchi [33] observed the gas-phase far-infrared absorption spectra of the NH_2, NHD, and ND_2 radicals in the 51 cm^{-1} to 366 cm^{-1} region with a high-resolution Fourier transform spectrometer. The NH_2 radical was generated in a multiple-traversal absorption cell by a DC discharge in an NH_3 and and Ar mixture. A discharge in an NH_3, D_2, and Ar mixture was used for production of NHD and ND_2. The observed spectra with a resolved fine structure were analyzed by Watson's A-reduced Hamiltonian, including a spin-rotation interaction term. The rotational and spin-rotation constants were determined to higher order. In the same experiment of NH_2, the rotational spectrum of the NH radical was also observed. From the rotational constants, molecular geometry parameters, and inertia defect, Δ, were determined as follows in Table 4.3.

Table 4.3: Molecular Geometry Parameters and Inertia Defect Δ for NH_2 and ND_2 in the Electronic Ground States

Parameter	NH_2	ND_2
r_0 (Å)^a	1.0245(37)	1.0239(19)
θ_0 (degrees)^a	103.34(51)	103.33(27)
Δ_{obs} (amu Å^2)^a	0.049561(4)	0.06770(4)
Δ_{calc} (amu Å^2)	0.047653	0.066566

^{a}One standard deviation limit in parentheses.

4.3.3 Spectrum Obtained by the Leone Group at JILA

One of the areas in which the most progress has been made has been the development of pulse-to-pulse normalization of time-resolved Fourier transform emission measurements in the near infrared. These are described by Lindner, Lundberg, Williams, and Leone [34], in *Review of Scientific Instruments*, 66:2812-2817, 1995. The spectra of NH_2 and its isotopologues can be difficult to unravel. This led R. W. Field and his colleagues, Coy and Jacobson, to propose methods for "Identifying patterns in multicomponent signals by extended cross correlation" [35]. These methods have now been widely used for studying fragmentation of ammonia and its isotopologues.

One of the first applications of these techniques is the study the photo-fragmentation of ammonia at 193.3 nm. Loomis, Reid, and Leone [36] showed that the fragmentation led to bimodal rotational distributions and to vibrational excitation of NH_2 ($\tilde{A}$). Photodissociation dynamics of $\tilde{A}$-state NH_3 and deuterated isotopologues had been investigated before by Mordaunt, Dixon, and Ashfold [37, 38] using the H(D) Rydberg atom photofragment translational spectroscopy, and a detailed model has been developed by Dixon [39].

A portion of the NH_2 emission spectrum obtained by Loomis, Reid, and Leone is shown in Figure 4.34. The transitions originate from the $N' = K_a'$ rotational states in $v_2' = 0$. The expected rotational structure, including the $N' > K_a'$ progressions and the doublets attributed to spin-rotation interaction, is also visible.

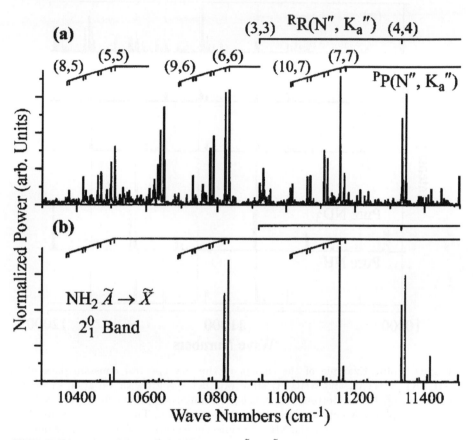

Figure 4.34: A small portion of the NH_2 $\tilde{A} \to \tilde{X}$ spectrum obtained using an InSb detector following the 193.3 nm photolysis of (a) room-temperature and (b) jet-cooled ammonia. Reproduced from Loomis et al., *J. Chem. Phys.* 112:658, 2000 [36], with the permission of AIP Publishing.

This work was followed by a more extensive application of the use of extended cross-correlation to examine the vibrational dynamics in the photofragmentation of NH_2D and ND_2H [40]. They also studied the competition between N-H and N-D bond cleavage in the photodissociation of NH_2D and ND_2H [41].

An example of the emission spectra obtained from an initial mixture of NH_3 and ND_3 is shown in Figure 4.35. Further deuterated isotopologues of ammonia are formed during equilibration of the initial mixture.

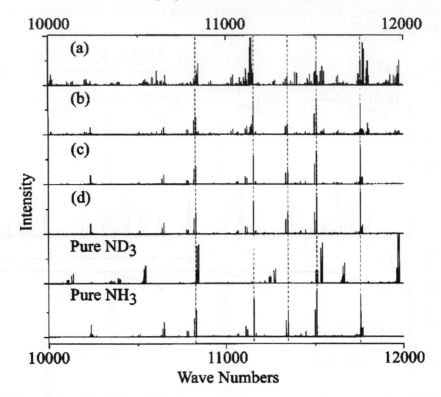

Figure 4.35: Example of the change in the raw spectral emission data of the $NH_2 (\tilde{A})$, $ND_2 (\tilde{A})$, and $NHD (\tilde{A})$ state product species with variation in mixture composition. In the bottom two panels, the emission spectra from pure NH_3 and pure ND_3 photodissociations are shown for comparison. The dashed lines are shown as a visual aid to examine the $NH_2 (\tilde{A})$ emission features. Neglecting error bars, the mixture compositions are approximately: (a) 10% NH_3, 40% NH_2D, 40% ND_2H, and 10% ND_3; (b) 25% NH_3, 50% NH_2D, 20% ND_2H, and 5% ND_3; (c) 55% NH_3, 25% NH_2D, 15% ND_2H, and <5% ND_3; (d) 80% NH_3, 15% NH_2D, <5% ND_2H, and <5% ND_3; Reproduced from Reid et al., *J. Chem. Phys.* 112:3181, 2000 [40], with the permission of AIP Publishing.

The following Figure 4.36 shows an example of product state distributions obtained with the spectral cross-correlation method.

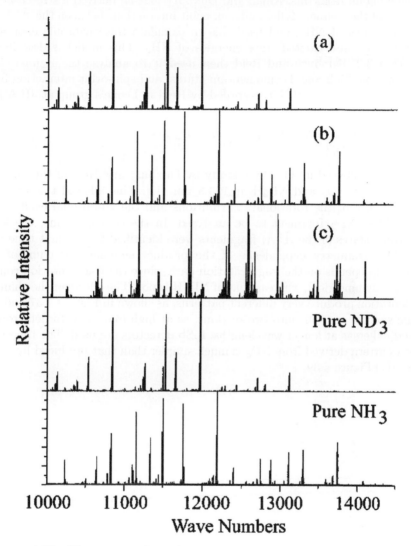

Figure 4.36: The extracted product-state spectral patterns from the room-temperature photodissociation of mixtures of NH_3, NH_2D, ND_2H and ND_3. The spectral patterns are from all of the accessed rotational states for the product vibrational states: (a) ND_2, $v_2 = 0$, (b) NH_2 $v_2 = 0, 1, 2$, (c) ND_2 $v_2 = 1, 2$. In the bottom two panels, the emission spectra of NH_2 ($\tilde{A}$) and ND_2 ($\tilde{A}$) from pure NH_3 and pure ND_3 photodissociations are shown for comparison. In all cases, the spectra are normalized so that the peak maximum is unity. Reproduced from Reid et al., *J. Chem. Phys.* 112:3181, 2000 [40], with the permission of AIP Publishing.

4.3.4 Stretch-Bender and Spin-Orbit Calculations

Following on from this, Alijah and Duxbury [42–44] derived a stretch-bender model of the Renner–Teller and spin-orbit interactions between the $\tilde{A}^2A_1$ and $\tilde{X}^2B_1$ states of NH_2, and used this to calculate the rovibronic energies in the ground and excited state energies of NH_2. This model is described in Section 3.2. Duxbury and Reid then used it to analyse the Renner–Teller interaction, high angular momentum states, and spin-orbit interaction in the electronic spectrum of ND_2 recorded by Reid in Leone's group at JILA [45].

4.3.5 Photodissociation Spectra of ND_3 and ND_2H Analysed by Duxbury and Reid

This was followed in 2011 by a study by Duxbury and Reid of the photodissociation of ND_3 and ND_2H at 193.3 nm [46]. This allowed the symmetry dependence of the rotational distributions and the vibrational excitation of the $\tilde{A}^2A_1$ ND_2 fragment to be analysed. In the course of this, the lowest vibronic states of the $\tilde{A}^2A_1$ fragments were identified for the first time [46].

The symmetry dependence of the product emission spectrum of ND_2 ($\tilde{A}^2A_1$) depends on the fragmentation path, since the fragmentation route in ND_3 is symmetrical, whereas that of ND_2H is not. The effects of the change in orientation produce very different patterns, as shown in Figures 4.37 and 4.38. The spectra are split into two sections, as at high frequency, Si detectors are used, whereas at longer wavelengths, InSb detectors are used. The fragmentation pattern derived from ND_3 is much simpler than that produced by ND_2H; see also Figure 4.39.

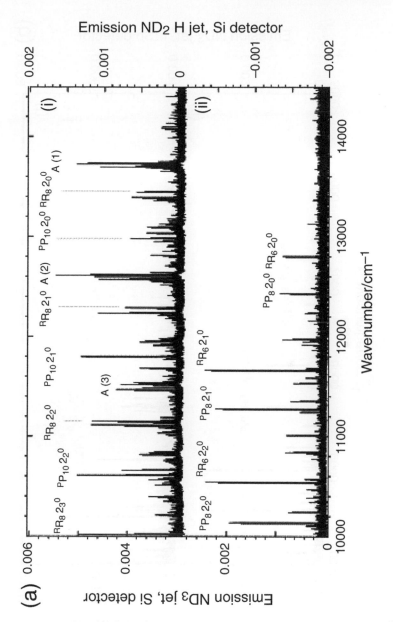

Figure 4.37: A survey of the entire product emission spectrum of ND_2 ($\tilde{A}^2A_1$) (i) ND_2H, and (ii) ND_3. Note the complexity of the emission spectrum of ND_2H and the simplicity of the emission spectrum of ND_3. Reprinted from Duxbury and Reid, *J. Mol. Spectrosc.* 267:123, 2011 [46]. Copyright (2011), with permission from Elsevier.

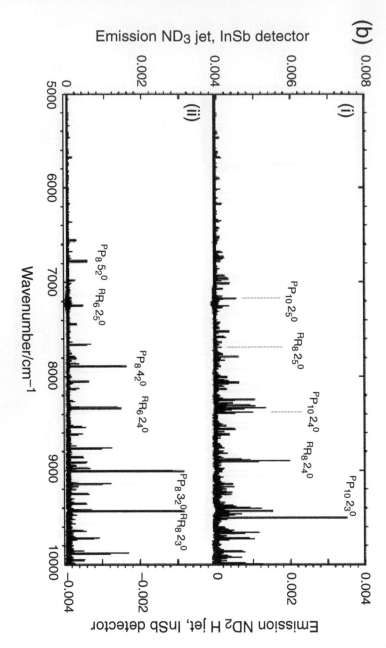

Figure 4.38: Continuation of Figure 4.37.

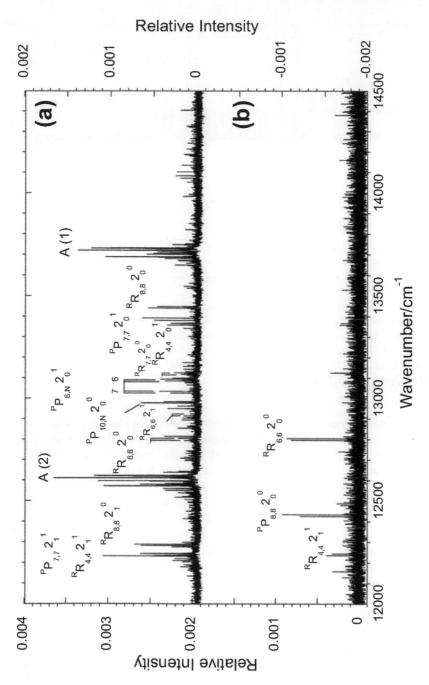

Figure 4.39: An expanded plot of the high wavenumber part, Figure 4.38, to show the complexity of the emission spectrum produced by jet-cooled ND_2H (a), and the simplicity and limited wavenumber range produced by the photolysis of ND_3 (b). Reprinted from Duxbury and Reid, *J. Mol. Spectrosc.* 267:123, 2011 [46]. Copyright (2011), with permission from Elsevier.

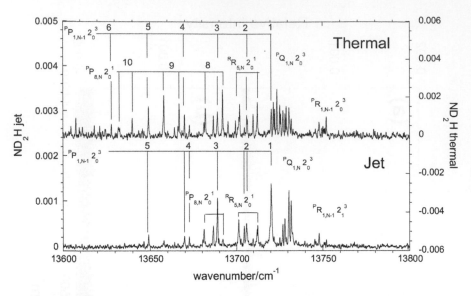

Figure 4.40: An expanded plot of the emission spectrum of region A(1) of the high wavenumber part of Figure 4.39 (a) to show the complexity of the emission spectrum produced by room-temperature photolysis (Thermal) and jet-cooled photolysis (Jet) of ND_2H. The three characteristic branches of the Σ subbands with $K'_a \equiv K' = 0$ are $^PR_{1,N-1}2_0^3$, $^PQ_{1,N}2_0^3$, and $^PP_{1,N-1}2_0^3$. Reprinted from Duxbury and Reid, *J. Mol. Spectrosc.* 267:123, 2011 [46]. Copyright (2011), with permission from Elsevier.

A detailed analysis of the Renner–Teller effect in the spectrum of NH_2 and ND_2 was given by Dressler and Ramsay in 1959 [27]. In their analysis, they were able to reach the $(0, 10, 0)$ level of NH_2, but only the $(0, 9, 0)$ level of ND_2. However, clear examples of the strongest Σ vibronic subbands could be seen.

In Figure 4.40, an expanded plot of the emission spectrum of ND_2 is shown, in which the three characteristic branches of the Σ subband may be seen. These are very similar to the bands displayed in Figure 4 of the paper by Dressler and Ramsay [27]. The main difference is that with newer instrumentation it is possible to access the lowest branches, and also to compare the spectra recorded under Thermal and Jet conditions. It can also be shown by looking at Figure 4.39, that the Σ subbands can only be identified in the emission spectra of ND_2H.

In addition to Σ subbands, there are also the $^RR_{5,N}2_0^1$ and $^PP_{8,N}2_0^1$ subbands. Note the Σ subband shows significant rotational structure, even when derived from the jet-cooled parent, whereas the rotational structure of $^PP_{8,N}2_0^1$ is only well developed in room-temperature photolysis (the Thermal group).

The next plot is of the emission spectrum of region A(2), Figure 4.41. In addition to the information about the region of the emission spectra, there

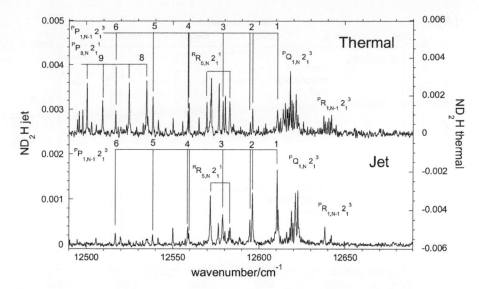

Figure 4.41: An expanded plot of the emission spectrum of region A(2), from room-temperature photolysis (Thermal) and jet-cooled photolysis (Jet) of ND_2H. Note the similarity of the bands to Figure 4.40, and the absence of the $^P P_{8,N} 2_1^1$ subband in the jet-cooled spectra. Jet-cooled ND_2H (a), and the simplicity and limited wavenumber range produced by the photolysis of ND_3. Reprinted from Duxbury and Reid, *J. Mol. Spectrosc.* 267:123, 2011 [46]. Copyright (2011), with permission from Elsevier.

is also evidence for the role played by Fermi resonance interactions between higher lying vibrational states in the $\tilde{X}^2 B_1$ state.

In Figures 4.42 (a) and (b), the effects of Fermi resonance are compared for emission from a jet-cooled gas (a), and a room-temperature gas (b). The first Fermi resonance in the $\tilde{X}^2 B_1$ state is between $(0, 2, 0)$ and $(1, 0, 0)$, and the second between $(0, 3, 0)$ and $(1, 1, 0)$. It may be seen that the rotational spacings within each series are almost identical. Additional structure appears on the long wavelength side of $^P P_{10,N} 2_{v_2''}^0$ when v_2'' is greater than 1. We associate this structure with Fermi-resonance interactions between the higher lying vibrational states in the $\tilde{X}^2 B_1$ state. In (a), the spectra in the regions of $^P P_{10,N} 2_2^0$ and $^P P_{10,N} 2_3^0$ are contrasted.

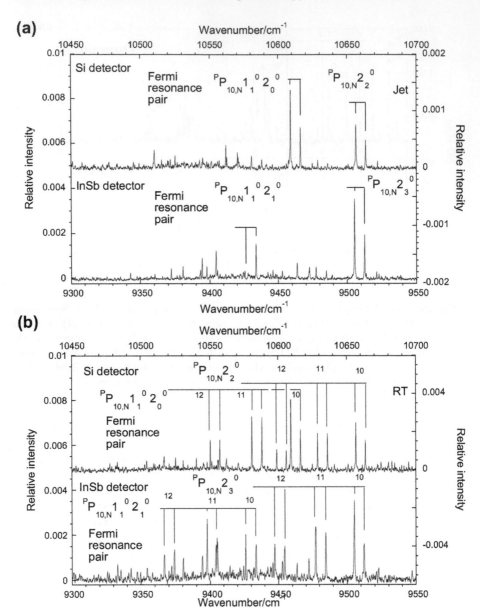

Figure 4.42: Effects of Fermi resonance for emission from (a) a jet-cooled gas (upper figure) and (b) a room-temperature gas (lower figure). Reprinted from Duxbury and Reid, *J. Mol. Spectrosc.* 267:123, 2011 [46]. Copyright (2011), with permission from Elsevier.

Further evidence of vibrational resonances is shown in Figure 4.43 (c) Jet and (d) Thermal, in which $^P P_{10,N} 2^0_2$ and $^P P_{10,N} 2^0_4$ are contrasted. Two extra subbands are now observed to the red of $^P P_{10,N} 2^0_4$. The three subbands are attributed to the three possible components of the triad, $^{P}P_{10,N}2_{4}^{0}$, $^{P}P_{10,N}1_{2}^{0}2_{2}^{0}$, and $^{P}P_{10,N}1_{2}^{0}2_{0}^{0}$.

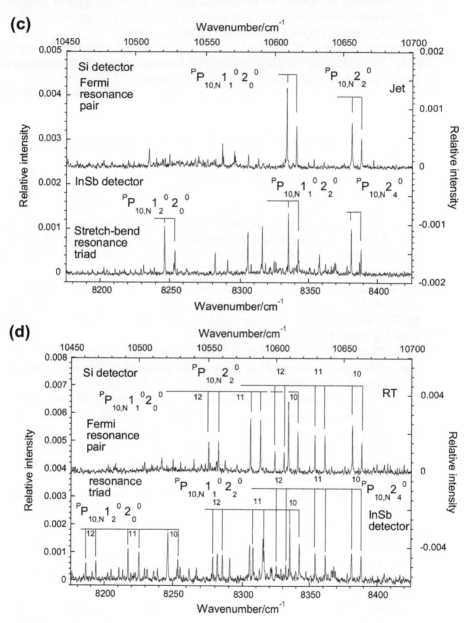

Figure 4.43: Vibrational resonances from (c) jet-cooled gas and (d) thermal gas. Reprinted from Duxbury and Reid, *J. Mol. Spectrosc.* 267:123, 2011 [46]. Copyright (2011), with permission from Elsevier.

4.4 Bibliography

[1] D. Feldmann, K. Meier, R. Schmiedl, and K. H. Welge. Laser induced fluorescence spectrum of singlet methylene. Spectroscopy of the $\tilde{a}^1A_1$ state. *Chemical Physics Letters*, 60(1):30–35, 1978.

[2] G. Herzberg and J. W. C. Johns. The spectrum and structure of singlet CH_2. *Proceedings of the Royal Society of London. Series A. Mathematical and Physical Sciences*, 295(1441):107–128, 1966.

[3] A. R. W. McKellar, P. R. Bunker, T. J. Sears, K. M. Evenson, R. J. Saykally, and S. R. Langhoff. Far infrared laser magnetic resonance of singlet methylene: Singlet-triplet perturbations, singlet-triplet transitions, and the singlet-triplet splittings. *The Journal of Chemical Physics*, 79(11):5251–5264, 1983.

[4] H. Petek, D. J. Nesbitt, P. R. Ogilby, and C. Bradley Moore. Infrared flash kinetic spectroscopy: The ν_1 and ν_3 spectra of singlet methylene. *The Journal of Physical Chemistry*, 87(26):5367–5371, 1983.

[5] H. Petek, D. J. Nesbitt, C. Bradley Moore, F. W. Birss, and D. A. Ramsay. Visible absorption and magnetic-rotation spectroscopy of $^1CH_2$: Analysis of the $^1A_1$ state and the $^1A_1 - {}^3B_1$ coupling. *The Journal of Chemical Physics*, 86(3):1189–1205, 1987.

[6] G. V. Hartland, W. Xie, D. Qin, and H.-L. Dai. Strong asymmetry induced $\Delta K_a = 3$ transitions in the CH_2 $\tilde{b}^1B_1 - \tilde{a}^1A_1$ spectrum: A study by Fourier transform emission spectroscopy. *The Journal of Chemical Physics*, 97(9):7010–7012, 1992.

[7] G. V. Hartland, D. Qin, and H.-L. Dai. Renner–Teller effect on the highly excited bending levels of $\tilde{a}^1A_1$ CH_2. *The Journal of Chemical Physics*, 102(17):6641–6645, 1995.

[8] H. Petek, D. J. Nesbitt, D. C. Darwin, and C. Bradley Moore. Visible absorption and magnetic-rotation spectroscopy of $^1CH_2$: The analysis of the $\tilde{b}^1B_1$ state. *The Journal of Chemical Physics*, 86(3):1172–1188, 1987.

[9] W. H. Green, N. C. Handy, P. J. Knowles, and S. Carter. Theoretical assignment of the visible spectrum of singlet methylene. *The Journal of Chemical Physics*, 94(1):118–132, 1991.

[10] I. García-Moreno, E. R. Lovejoy, C. Bradley Moore, and G. Duxbury. Radiative lifetimes of CH_2 (b^1B_1). *The Journal of Chemical Physics*, 98(2):873–882, 1993.

[11] G. V. Hartland, D. Qin, and H.-L. Dai. Fourier transform dispersed fluorescence spectroscopy: Observation of new vibrational levels in the 5000–8000 cm^{-1} region of $\tilde{a}^1A_1$ CH_2. *The Journal of Chemical Physics*, 98(3):2469–2472, 1993.

[12] I. García-Moreno and C. Bradley Moore. Spectroscopy of methylene: Einstein coefficients for CH_2 $(\tilde{b}^1B_1 - \tilde{a}^1A_1)$ transitions. *The Journal of Chemical Physics*, 99(9):6429–6435, 1993.

[13] B.-C. Chang, M. Wu, G. E. Hall, and T. J. Sears. Near-infrared vibronic spectrum of the CH_2 $\tilde{b}^1B_1 \leftarrow \tilde{a}^1A_1$ transition. *The Journal of Chemical*

Physics, 101(11):9236–9245, 1994.

[14] J. C. Bloch, R. W. Field, G. E. Hall, and T. J. Sears. Time-resolved frequency modulation spectroscopy of photochemical transients. *The Journal of Chemical Physics*, 101(2):1717–1720, 1994.

[15] C. Fockenberg, A. J. Marr, T. J. Sears, and B.-C. Chang. Near-infrared high resolution diode laser spectrum of the CH_2 $\tilde{b}\,^1B_1 \leftarrow \tilde{a}^1A_1$ transition. *Journal of Molecular Spectroscopy*, 187(2):119–125, 1998.

[16] J.-P. Gu, G. Hirsch, R. J. Buenker, M. Brumm, G. Osmann, P. R. Bunker, and P. Jensen. A theoretical study of the absorption spectrum of singlet CH_2. *Journal of Molecular Structure*, 517–518:247–264, 2000.

[17] P. Bunker and P. Jensen. *Molecular Symmetry and Spectroscopy*. Canadian Science Publishing (NRC Research Press), 2nd revised edition edition, January 2006.

[18] G. Duxbury, B. D. McDonald, M. Van Gogh, A. Alijah, Ch. Jungen, and H. Palivan. The effects of vibrational resonances on Renner–Teller coupling in triatomic molecules: The stretch-bender approach. *The Journal of Chemical Physics*, 108(6):2336–2350, 1998.

[19] G. Duxbury, A. Alijah, B. D. McDonald, and Ch. Jungen. Stretch-bender calculations of the effects of orbital angular momentum and vibrational resonances in the spectrum of singlet methylene. *The Journal of Chemical Physics*, 108(6):2351–2360, 1998.

[20] A. J. Marr, T. J. Sears, and B.-C. Chang. Near-infrared spectroscopy of CH_2 by frequency modulated diode laser absorption. *The Journal of Chemical Physics*, 109(9):3431–3442, 1998.

[21] G. Duxbury and Ch. Jungen. Effects of orbital angular momentum in CH_2: The Renner–Teller effect. *Molecular Physics*, 63(6):981–998, 1988.

[22] K. Kobayashi, L. D. Pride, and T. J. Sears. Absorption spectroscopy of singlet CH_2 near 9500 cm^{-1}. *The Journal of Physical Chemistry A*, 104(45):10119–10124, 2000.

[23] C.-H. Chang, G. E. Hall, and T. J. Sears. Sub-Doppler spectroscopy of mixed state levels in CH_2. *The Journal of Chemical Physics*, 133(14):144310, 2010.

[24] C.-H. Chang, Z. Wang, G. E. Hall, T. J. Sears, and J. Xin. Transient laser absorption spectroscopy of CH_2 near 780 nm. *Journal of Molecular Spectroscopy*, 267(1–2):50–57, 2011.

[25] A. Kalemos, T. H. Dunning Jr., A. Mavridis, and J. F. Harrison. CH_2 revisited. *Canadian Journal of Chemistry*, 82(6):684–693, 2004.

[26] C.-H. Chang, J. Xin, T. Latsha, E. Otruba, Z. Wang, G. E. Hall, T. J. Sears, and B.-C. Chang. CH_2 $\tilde{b}\,^1B_1 - \tilde{a}^1A_1$ band origin at 1.20 μm. *The Journal of Physical Chemistry A*, 115(34):9440–9446, 2011. PMID: 21314141.

[27] K. Dressler and D. A. Ramsay. The electronic absorption spectra of NH_2 and ND_2. *Philosophical Transactions of the Royal Society of London. Series A, Mathematical and Physical Sciences*, 251:553–602, 1959.

[28] J. W. C. Johns, D. A. Ramsay, and S. C. Ross. The $\tilde{A}^2A_1 - \tilde{X}^2B_1$

absorption spectrum of NH_2 between 6250 and 9500 Å. *Canadian Journal of Physics*, 54(17):1804–1814, 1976.

[29] R. E. Muenchausen, G. W. Hills, M. F. Merienne-Lafore, D. A. Ramsay, M. Vervloet, and F. W. Birss. Further studies of the electronic and vibrational spectra of ND_2: Improved molecular parameters for the 000 and 010 levels of the ground state. *Journal of Molecular Spectroscopy*, 112(1):203–210, 1985.

[30] M. Vervloet. Study of the electronic ground state of NH_2 by laser excited fluorescence Fourier transform spectroscopy. *Molecular Physics*, 63(3):433–449, 1988.

[31] R. N. Dixon, S. J. Irving, J. R. Nightingale, and M. Vervloet. Transitions between high angular momentum states in the electronic spectrum of NH_2. *J. Chem. Soc., Faraday Trans.*, 87:2121–2133, 1991.

[32] I. H. Bachir, T. R. Huet, J.-L. Destombes, and M. Vervloet. A combined analysis of laser optogalvanic and Fourier transform emission spectra of NH_2 near its barrier to linearity. *Journal of Molecular Spectroscopy*, 193(2):326–353, 1999.

[33] I. Morino and K. Kawaguchi. Fourier transform far-infrared spectroscopy of the NH_2, NHD, and ND_2 radicals. *Journal of Molecular Spectroscopy*, 182(2):428–438, 1997.

[34] J. Lindner, J. K. Lundberg, R. M. Williams, and S. R. Leone. Pulse-to-pulse normalisation of time-resolved Fourier transform emission experiments in the near infrared. *Review of Scientific Instruments*, 66:2812–2817, 1995.

[35] S. L. Coy, M. P. Jacobson, and R. W. Field. Identifying patterns in multicomponent signals by extended cross correlation. *The Journal of Chemical Physics*, 107(20):8357–8369, 1997.

[36] R. A. Loomis, J. P. Reid, and S. R. Leone. Photofragmentation of ammonia at 193.3 nm: Bimodal rotational distributions and vibrational excitation of NH_2 ($\tilde{A}$). *The Journal of Chemical Physics*, 112(2):658–669, 2000.

[37] D. H. Mordaunt, M. N. R. Ashfold, and R. N. Dixon. Photodissociation dynamics of $\tilde{A}$ state ammonia molecules. I. State dependent $\mu - v$ correlations in the NH_2 (ND_2) products. *The Journal of Chemical Physics*, 104(17):6460–6471, 1996.

[38] D. H. Mordaunt, R. N. Dixon, and M. N. R. Ashfold. Photodissociation dynamics of $\tilde{A}$ state ammonia molecules. II. The isotopic dependence for partially and fully deuterated isotopomers. *The Journal of Chemical Physics*, 104(17):6472–6481, 1996.

[39] R. N. Dixon. Photodissociation dynamics of $\tilde{A}$ state ammonia molecules. III. A three-dimensional time-dependent calculation using ab initio potential energy surfaces. *Molecular Physics*, 88(4):949–977, 1996.

[40] J. P. Reid, R. A. Loomis, and S. R. Leone. Characterization of dynamical product-state distributions by spectral extended cross-correlation: Vibrational dynamics in the photofragmentation of NH_2D and ND_2H.

The Journal of Chemical Physics, 112(7):3181–3191, 2000.

[41] J. P. Reid, R. A. Loomis, and S. R. Leone. Competition between N-H and N-D bond cleavage in the photodissociation of NH_2D and ND_2H. *Journal of Physical Chemistry A*, 104(45):10139–10149, 2000.

[42] A. Alijah and G. Duxbury. Renner–Teller and spin-orbit interactions between the $\tilde{A}^2A_1$ and the $\tilde{X}^2B_1$ states of NH_2: The stretch-bender approach. *Journal of Molecular Spectroscopy*, 211(1):7–15, 2002.

[43] A. Alijah and G. Duxbury. Stretch-bender calculations of the rovibronic energies in the $\tilde{X}^2B_1$ electronic ground state of NH_2. *Journal of Molecular Spectroscopy*, 211(1):16–30, 2002.

[44] G. Duxbury and A. Alijah. Stretch-bender calculations of the rovibronic energies in the excited, $\tilde{A}^2A_1$, electronic state of NH_2 and of the near-resonant high-lying levels of the $\tilde{X}^2B_1$ state. *Journal of Molecular Spectroscopy*, 211(1):31–57, 2002.

[45] G. Duxbury and J. P. Reid. Renner–Teller interaction, high angular momentum states and spin-orbit interaction in the electronic spectrum of ND_2. *Molecular Physics*, 105(11–12):1603–1618, 2007.

[46] G. Duxbury and J. P. Reid. Photodissociation of ND_3 and ND_2H at 193.3 nm: Symmetry dependence of the rotational distributions and vibrational excitation of the ND_2 fragment. *Journal of Molecular Spectroscopy*, 267(1–2):123–135, 2011.

CHAPTER 5

Second Row Dihydrides

CONTENTS

5.1 THE SPECTRUM OF SiH$_2$

5.1.1 Introduction

The silicon analogue of methylene, silylene SiH$_2$, was first detected by I. Dubois, G. Herzberg, and R. D. Verma in 1967 [1]. This allowed the main rotation constants A, B, and C to be derived for the $\tilde{A}\,^1B_1(0,1,0)$ and $\tilde{A}\,^1B_1(0,2,0)$ upper states as well as for the $\tilde{X}\,^1A_1(0,0,0)$ lower state.

Following on from this, Ivan Dubois [2] produced a high resolution spectrum of SiH$_2$ mixing silane (SiH$_4$) from Matheson of Canada Ltd with H$_2$ in

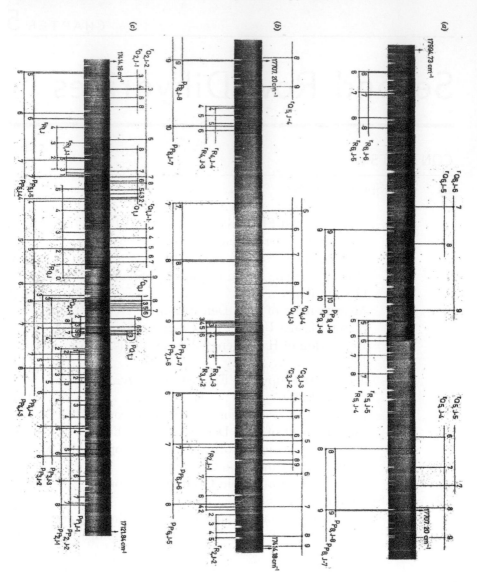

Figure 5.1: The complete (0,2,0)–(0,0,0) absorption band of SiH$_2$ near 5700 Å. Reproduced by permission from Dubois, *Canad. J. Phys.* 46:2485, 1968 [2]. © 1968 Canadian Science Publishing.

the ratio 1:50. The optimum pressure of this mixture in the absorption tube appeared to be 2 mm Hg, i.e., the partial pressure of SiH$_4$ was 40 μm Hg. The absorption was strongest for a delay between the source flash and the discharge of about 5 μs and disappeared after 20 μs.

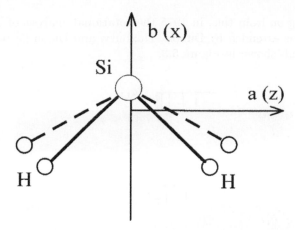

Figure 5.2: Equilibrium geometry of SiH₂ in the $\tilde{X}\,^1A_1$ (full line) and $\tilde{A}\,^1B_1$ (dotted line) electronic states. The bending angle is 92° in the ground state, but $\sim 120°$ in the excited state. Reproduced from Escribano and Campargue, *J. Chem. Phys.*, 108:6249, 1998 [3], with the permission of AIP Publishing.

The best spectrum of the (0,2,0)–(0,0,0) absorption band, shown in Figure 5.1, was obtained with a path length of 48 m, i.e., 24 traversals of a 2 m absorption tube. The measurements were made in the second order of a 10 m Eagle spectrograph using a 600 line/nN grating, where n is the order and N is the total number of lines. The plates were measured on a photographic comparator using Fe and Ne lines from a hollow cathode discharge as a reference spectrum. The relative accuracy of the well-defined lines was estimated to be ca. ± 0.01 cm^{-1}.

The spectrum, which consists of a progression of three bands separated by about 850 cm^{-1}, shows a widely open rotational structure due to the large geometry change, shown in Figure 5.2, from the ground $\tilde{X}\,^1A_1(0,0,0)$ state to the $\tilde{A}\,^1B_1(0,v_2',0)$ progression from $\tilde{A}\,^1B_1(0,1,0)$ to $\tilde{A}\,^1B_1(0,3,0)$. Dubois [2] carried out vibrational and rotational analyses of the absorption spectrum of SiH₂. This led to the determination of the rotational constants in the lower $^1A_1$ state and those of the (0,1,0), (0,2,0), and (0,3,0) levels of the upper $^1B_1$ state, allowing an extrapolation to the position of the (0,0,0) state.

Tables 1 to 3 were deposited in the "Depository for Unpublished Data" in the "National Science Library of Ottawa, Canada" [4]. The copies used here were provided by Alain Campargue of the Laboratoire de Spectrométrie Physique, (UMR 5588), at the Université Joseph Fourier de Grenoble, France.

When these spectra were examined in detail by Dubois [2], it was found that 1586 combination differences could be identified in the ground state with a standard deviation of 0.03 cm^{-1}. Rather different behaviour was identified in the upper $^1B_1$ state. In his analysis, it was found that the standard deviation for the three $^1B_1$ levels could not be reduced below 1.15 cm^{-1}.

Following on from this, in 1975 the rotational analysis of the absorption spectrum was extended by Dubois, Duxbury, and Dixon [5] to the 4–0, 5–0, and 6–0 bands shown in Figure 5.3.

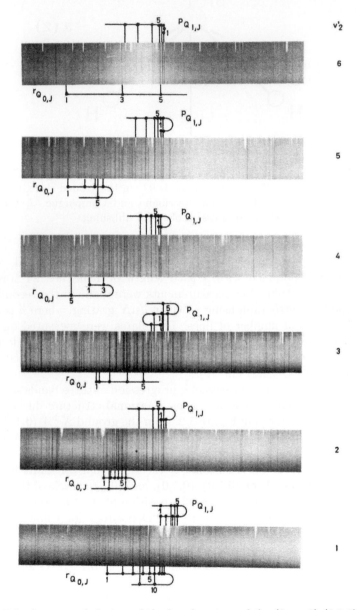

Figure 5.3: An expanded view of the band centres of the $(0, v_2, 0)$-$(0,0,0)$ progression in the absorption spectrum of SiH_2 from $v_2 = 1$ to $v_2 = 6$. Reproduced from Dubois et al., *J. Chem. Soc., Faraday Trans. 2* 71:799, 1975 [5], with permission from The Royal Society of Chemistry.

The two combining states, the $\tilde{X}\,^1A_1$ and the $\tilde{A}\,^1B_1$ states, were shown to be derived from a $^1\Delta$ electronic state split by static Renner–Teller interaction. The vibronic pattern of the excited state was shown to be due both to large amplitude vibration, and also to the Renner–Teller effect. The barrier to linearity in the $\tilde{A}\,^1B_1$ state was estimated to lie at ca. $8000\,\text{cm}^{-1}$. Unlike the spectrum of CH_2 recorded by Alijah and Duxbury [6], the position of the origin of the $^3B_1$ state of SiH_2 was calculated to lie above the $\tilde{X}\,^1A_1$ ground state. The 4–0, 5–0, and 6–0 bands are even more perturbed than those of the (0,1,0), (0,2,0), and (0,3,0) bands, and predissociation became much more noticeable for $v_2 = 5$ and 6.

5.1.2 Laser-Excited Fluorescence Detection of SiH_2

In the experiments of J. W. Thoman Jr and J. I. Steinfeld [7], organosilanes were dissociated by infrared multiple-photon dissociation (IR-MPD), and the resulting SiH_2 detected by laser-induced fluorescence (LIF). Phenylsilane and n-butylsilane were degassed. A slow flow of gas at 5 to 20 mTorr pressure (1 Torr $\approx$ 133.3 Pa) was established by pumping on the liquid silanes at room temperature with a liquid nitrogen trapped turbomolecular pump. Pressures were controlled with a leak valve.

The laser used was a line tuneable transversely excited atmospheric (TEA) laser, which provides a 120 ns full width at half maximum (FWHM) pulse with a 1 μs tail. Infrared fluences in the range from 0.5–5 J/cm^2 obtained by focussing 5–50 mJ pulses with a 500 mm focal length BaF_2. The visible-wavelength probe beam, collinear, and counter-propagating to the CO_2 laser, probed only the central portion of the photolysis region. An attenuated XeCl-excimer pumped dye laser with rhodamine 6G dye produced pulses with 50 μJ in 10 ns FWHM, and 0.5 cm^{-1} spectral width. This excited transitions in the SiH_2 $\tilde{A}\,^1B_1(0,2,0) \rightarrow \tilde{X}\,^1A_1(0,0,0)$ vibronic band. A wideband pulse amplifier and a gated integrator/boxcar averager were used to collect the signal. A fast UV photodiode, which detected scattered light from the excimer laser, was used to trigger the boxcar averager. Because the trigger pulse was produced only a few ns before the LIF signal, it was necessary to delay the signal by using a 100 foot long coaxial cable. The photolysis and probe lasers could be synchronised to within ±100 ns.

SiH_2 was observed as a product in the infrared multiphoton decomposition of both n-butylsilane and phenylsilane under collision-free conditions (10–25 mTorr), or with added buffer gas (0.4–0.8% He). The LIF signals from phenylsilane are a factor of 30–50 weaker than with butylsilane, indicating that significantly less SiH_2 is produced in the former system. Since the infrared multiphoton decomposition of phenylsilane is much more facile than for butylsilane, or other n-alkylsilanes, the production of SiH_2, via SiH_4, must be a relatively minor channel in the decomposition of the phenyl system.

In the first figure, Figure 5.4, the LIF excitation spectrum of the (0,2,0)–(0,0,0) band of the $\tilde{A}\,^1B_1 - \tilde{X}\,^1A_1$ transition of SiH_2 is shown. The spectra of SiH$_2$ derived from butyl- and phenylsilane are quite similar, indicating equivalent product rotational-state distributions in the two cases.

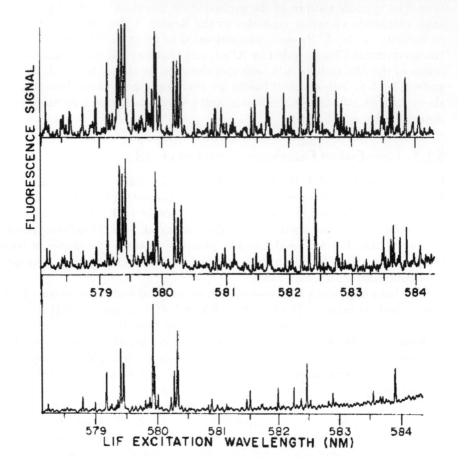

Figure 5.4: LIF excitation spectra of SiH$_2$ in the (0,2,0)–(0,0,0) band of the $\tilde{A}^1 B_1 - \tilde{X}^1 A_1$ transition derived from butyl- and phenylsilane are quite similar, indicating equivalent product rotational-state distributions in the two cases. Upper trace: 15 mTorr n-butylsilane, CO$_2$ 10 P(20) line. Middle trace: 25 mTorr phenylsilane, CO$_2$ 10P (32) line. Both spectra are recorded at "zero" delay between the infrared and visible laser pulse, and with fluorescence detection gate set to 0–50 ns following excitation. Bottom trace: 20 mTorr n-butylsilane in 400 Torr He, probed 36 μs after the photolysis laser, and with the sensitivity increased by a factor of 25. Reprinted from Thoman and Steinfeld, *Chem. Phys. Lett.* 124:35, 1986 [7]. Copyright (1986), with permission from Elsevier.

In the expanded section of LIF excitation spectrum of SiH$_2$, Figure 5.5, derived from Figure 5.4, the upper trace the detector gate is set to 0–50 ns following the excitation pulse. When the gate is set to a decay of 60–260 ns with

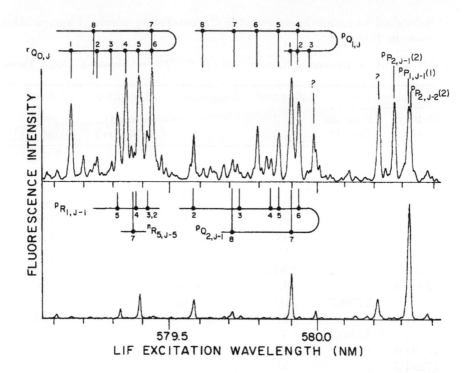

Figure 5.5: Expanded section of the LIF excitation spectrum of SiH_2 derived from 10 mTorr n-butylsilane, measured near the signal peak. Assignments are based on the data of Dubois. Upper trace: detector gate set to 0–50 ns following excitation pulse. Lower trace: detector gate set to 60–260 ns following the excitation pulse. Reprinted from Thoman and Steinfeld, *Chem. Phys. Lett.* 124:35, 1986 [7]. Copyright (1986), with permission from Elsevier.

respect to the pulse, most of the features are absent. However a few features remain in the spectrum, corresponding to anomalously long-lived excited rovibronic states. This can be clearly seen in Figure 5.5. Most of these long-lived features can be assigned to specific rotational levels of the $\tilde{A}\,^1B_1$ state, but there is no clear pattern of the J' values of these levels. Many unassigned lines were noted by Dubois in his earlier studies [2, 5]

Following on from the study by Thoman and Steinfeld [7], the measurements were extended to include wide fluctuations in fluorescence lifetimes of individual rovibronic levels in SiH_2 $\tilde{A}\,^1B_1$, by Thoman, Steinfeld, McKay, and Knight [8]. This was followed by a direct observation of silicon (^3P) following state-selected photofragmentation of $\tilde{A}\,^1B_1$ silylenes [9] which included the use of photofragmentation excitation spectroscopy (PHOFEX). Variations in Si atom production are measured simultaneously with the SiH_2, allowing comparisons to be made between Si yield and the rovibronic structure in the SiH_2 $^1B_1$ manifold. A comparison between wide fluctuations in fluorescent lifetimes of individual rovibronic levels and of Si (^3P) and state selected fragmentation is made in Table 5.1.

Table 5.1: A Comparison of Relative Yield of Si (^3P) and Measured Decay Times

Energy (vac)		Lifetimes[a]	Assignment	Absorption	Silicon
McKay (cm^{-1})	Thoman (cm^{-1})	(ns)		intensity (arbitrary)	yield (arbitrary)
	17273.9	8	$5_{15} - 4_{23}$		
17273.7		M	$7_{16} - 6_{24}$ $+6_{15} - 5_{23}$	5.97	172
17272.8		8			300
17271.3		196[b]			145
17270.1		108[b]			128
17269.3		M			41
17268.3		290[b]			162
17267.3		L	$2_{21} - 3_{31}$	2.05	49
	17266.6	16	$1_{11} - 1_{01}$		
17265.9		24	$2_{20} - 3_{30}$	6.11	133
	17265.4	7	$2_{20} - 3_{30}$		
17264.9		M			178
17264.5		M			113
17263.9		S			95
	17263.8	13	$2_{12} - 2_{02}$ $+$others		
17263.6		15	$2_{12} - 2_{02}$ $+5_{32} - 6_{42}$	4.19	165
17262.8		S			110
17262.0		150[b]	$3_{13} - 3_{03}$	12.76	227
	17261.8	150	$3_{13} - 3_{03}$		
	17261.8	7	$6_{06} - 5_{14}$		
17261.4		7[b]	$6_{06} - 5_{14}$	1.51	331
	17261.0	9	$4_{14} - 4_{04}$		
17260.9		L			254
17260.4		17	$4_{14} - 4_{04}$	5.07	345
17259.7		109[b]	$8_{26} - 7_{52}$	2.43	377
17259.2		109[b]	$5_{15} - 5_{05}$ $+5_{05} - 4_{13}$	17.25	5.14
	17258.6	12	$4_{04} - 3_{12}$ $+3_{03} - 2_{11}$		
17258.4		12[b]	$4_{04} - 3_{12}$ $+3_{03} - 2_{11}$	3.14	196
	17258.1	60	$6_{16} - 6_{06}$ $+7_{17} - 7_{07}$		

Table 5.1: A Comparison of Relative Yield of Si (^3P) and Measured Decay Times

Energy (vac)		Lifetimes[a]	Assignment	Absorption	Silicon
McKay (cm^{-1})	Thoman (cm^{-1})	(ns)		intensity (arbitrary)	yield (arbitrary)
17257.9		60^b	$6_{16} - 6_{06}$ $+7_{17} - 7_{07}$	22.49	245
17257.3		M			255
17257.0		M			134
17256.0		M			155
17255.5		S			151
	17254.2	137	$2_{11} - 2_{21}$		
	17254.2	8	$2_{11} - 2_{21}$		
17254.1		137^b	$2_{11} - 2_{21}$	2.21	207
17253.2		30^b	$8_{08} - 8_{18}$	14.02	244
17252.3		M			178
	17251.1	112			242
	17250.3	738	$8_{17} - 8_{27}$		
	17250.3	11	$8_{17} - 8_{27}$		
17250.2		11^b	$8_{17} - 8_{27}$	4.17	375
17249.9		7^b	$6_{34} - 7_{44}$ $+3_{12} - 3_{22}$	2.53	176
	17249.8	7			
	17249.5	106	$7_{07} - 7_{17}$		
	17249.5	28	$7_{07} - 7_{17}$		
17249.4		M			143
17248.7		L			110
	17248.6	23			
17247.9		22	$6_{06} - 6_{16}$	13.70	441
	17247.8	6	$6_{06} - 6_{16}$		
17247.4		S	$6_{15} - 5_{41}$	1.62	274
17246.9		L	$3_{22} - 4_{32}$	6.07	219
17246.5		S	$4_{13} - 4_{23}$	4.39	178
17245.7		10^b	$5_{14} - 5_{24}$ $+5_{05} - 5_{15}$	5.79	147
	17245.6	10	$5_{05} - 5_{15}$ $+5_{14} - 5_{24}$		
17244.3		151	$+2_{02} - 2_{12}$ $+7_{16} - 5_{26}$ $+1_{01} - 1_{11}$	10.03	165
	17244.2	121	$7_{16} - 7_{26}$ $+1_{01} - 1_{11}$		
17243.5		19^b	$3_{03} - 3_{13}$ $+4_{04} - 4_{14}$	19.02	329

Table 5.1: A Comparison of Relative Yield of Si (^3P) and Measured Decay Times

McKay (cm^{-1})	Thoman (cm^{-1})	Lifetimes[a] (ns)	Assignment	Absorption intensity (arbitrary)	Silicon yield (arbitrary)
	17243.2	19	$+6_{15} - 6_{25}$ $2_{02} - 2_{12}$ +others		

A comparison of relative yield of Si (^3P), McKay et al. [9], for various single rovibronic transitions in the $\tilde{A}\,^1B_1(0,2,0) \rightarrow \tilde{X}\,^1A_1(0,0,0)$ band of SiH$_2$ and the measured decay times for excitation features in the same band of SiH$_2$, Thoman et al. [8]. In order to compare the two sets of measurements in the table, both series are plotted from high frequency to low frequency.

[a]S: short lifetime range ($\tau \leq 5$ ns), M: medium range (5 ns $< \tau \leq 150$ ns), L: long range ($\tau > 150$ ns).

[b]Lifetimes from Thoman et al. (Reference [8] and unpublished).

Single rotational levels in various $\tilde{A}\,^1B_1(0,v_2',0)$ excited vibronic states of SiH$_2$ were prepared via laser excitation from the $\tilde{X}\,^1A_1$ ground state, using a Molectron ND:YAG 532 nm/355 nm pumped Molectron Dye laser. A variety of dyes was used. A second dye laser was frequency doubled and Raman shifted. This allowed the production of ultraviolet light resonant with the $3p^2(^3P_0)(J''=0) \rightarrow 4s(^3P_0^1)(J'=1)$ electronic transition of Si (251.507 nm). A schematic diagram of the PHOFEX spectroscopy instrument is shown in Figure 5.6.

Triggering of the three lasers, and all the detection equipment, was effected by a specially made programmable delay generator. The SiH$_2$ and Si probe lasers (10 Hz) were triggered to appear in the reaction region, coincident in time, and delayed after the IR-MPD laser, (5 Hz). This optimises the amount of intensity of the signal produced by a desired process. For example, Si produced as a result of the vibronic excitation of SiH$_2$ is itself produced from n-butylsilane via IR-MPD. The detection electronics were triggered to take advantage of the interleaved 5 Hz pulse rate of the IR-MPD CO$_2$ laser. This is achieved by subtracting the background Si and SiH$_2$ signals, measured without the IR-MPD laser firing, from the alternating signal derived from the three-laser process.

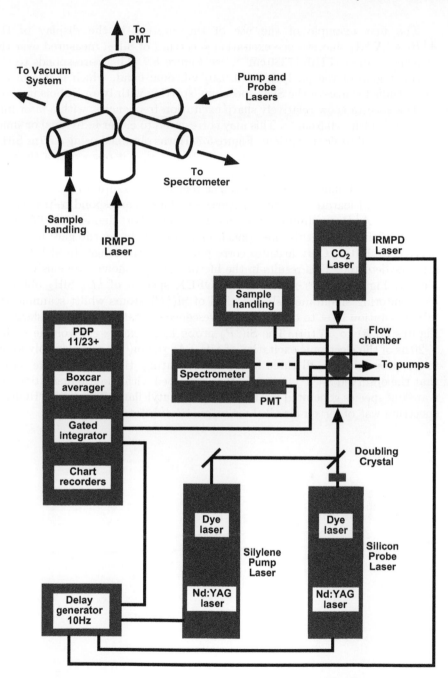

Figure 5.6: Schematic of experimental apparatus for PHOFEX spectroscopy. Inset: The flow chamber, showing laser orientations. Reproduced from McKay et al., *J. Chem. Phys.* 95:1688, 1991 [9], with the permission of AIP Publishing.

The first example of the use of this system is the display of the $\tilde{A}\,^1B_1 \leftarrow \tilde{X}\,^1A_1$ fluorescence excitation spectrum of SiH_2, measured over the frequency range 17210–17280 cm^{-1}; see Figure 5.7. This corresponds to the central region of the $(0,2,0) \rightarrow (0,0,0)$ vibronic band, which is one of the most studied ranges of the SiH_2 spectrum, starting with Iwan Dubois in 1968.

The spectra show relatively sharp excitation frequencies, with a laser limited bandwidth ≈ 0.5 cm^{-1}. This may correspond to either individual or small groups of rovibronic transitions. Figure 5.7 (i) was obtained using a 50 ns SiH_2. The fluorescence detection gate was adjusted so that it opened at a time coincident with the laser excitation of SiH_2. In Figure 5.7 (ii), the fluorescence excitation spectrum is delayed by ≈ 5 ns relative to Figure 5.7 (i). The intensity of several features diminishes markedly. These correspond to transitions that access $\tilde{A}\,^1B_1$ rovibronic levels that have very short fluorescence lifetimes. Figure 5.7 (iii) shows this effect much more markedly, as the gate had been delayed by 150 ns. Only features corresponding to the longer-lived $^1B_1$ state are now detected. This results in the lifetime ranges shown previously in Table 5.1. Figure 5.7 (iv) shows the PHOFEX spectra of $^1B_1$ SiH_2 obtained by monitoring simultaneously the yield of $Si(^3P)$ atoms whilst scanning the SiH_2 excitation laser to obtain the fluorescence excitation spectrum shown in Figures 5.7 (i) to 5.7 (iii). The $Si(^3P)$ probe laser excites the volume within ≤ 20 ns of the electronic excitation of SiH_2. As the $Si(^3P)$ signal is only found to be present when all three lasers are operating, the authors were happy that the observed $Si(^3P)$ atom yield is associated with the fragmentation of a transient species generated via IR-MPD of n-butylsilane. The same PHOFEX spectrum was observed for all three magnetic substates of $Si(^3P)$.

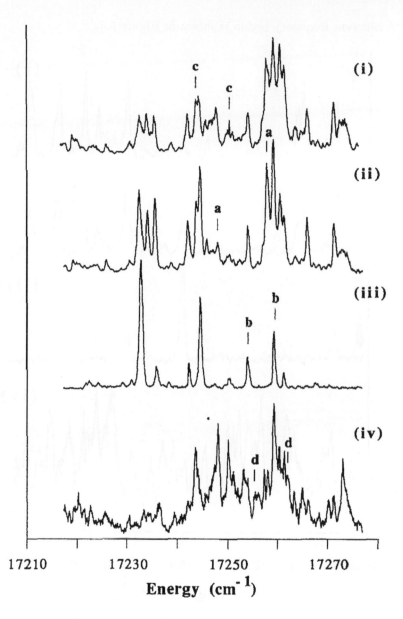

Figure 5.7: The $\tilde{A}\,^1B_1(0,2,0) \leftarrow \tilde{X}\,^1A_1(0,0,0)$ fluorescent excitation spectrum of SiH$_2$ measured in time: (i) t_0; (ii) $t_0 + 5$ ns; (iii) $t_0 + 150$ ns; and (iv) the PHOFEX spectrum, showing the relative yield of Si(^3P) as a function of the SiH$_2$ excitation frequency. Reproduced from McKay et al., *J. Chem. Phys.* 95:1688, 1991 [9], with the permission of AIP Publishing.

A different frequency region is shown in Figure 5.8.

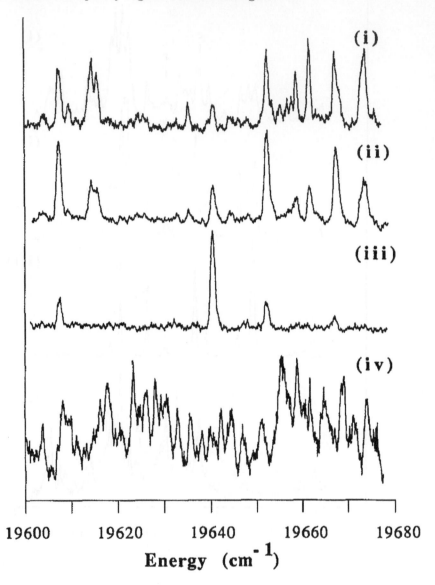

Figure 5.8: Data associated with an $80\,\mathrm{cm}^{-1}$ region of the $(0,5,0) \to (0,0,0)$ band, using a similar presentation method to that for $(0,2,0) \to (0,0,0)$. The analysis method used is equivalent to that of Figure 5.7. Reproduced from McKay et al., *J. Chem. Phys.* 95:1688, 1991 [9], with the permission of AIP Publishing.

McKay and her colleagues [9] have tried to establish the relationship between the photodissociation of SiH_2 and the appearance of $Si(^3P)$. Their main route was to determine whether or not there is a correlation between the $Si(^3P)$ yield and the SiH_2 lifetime. The method that they used was by making a scatterplot. They showed that the correlation between $Si(^3P)$ yield and absorption intensity appeared to be stronger than that between the $Si(^3P)$ yield and the SiH_2 lifetime.

The evidence for a direct correlation between $Si(^3P)$ yield and SiH_2 $\tilde{A}\,^1B_1 \leftarrow \tilde{X}\,^1A_1$ was explored using data available from Jasinski [10, 11]. Using this, it was possible to compare the central region of the $\tilde{A}\,^1B_1(0,2,0) \leftarrow \tilde{X}\,^1A_1(0,0,0)$ PHOFEX spectrum with a high resolution, (Doppler-limited linewidth $\approx 40\,MHz$), with the transient absorption spectrum of SiH_2, produced by the $193\,nm$ photolysis of phenylsilane.

They found that every feature present in the $Si(^3P)$ PHOFEX spectrum may be correlated with a corresponding feature in the Jasinski absorption spectrum. This higher-resolution scan reveals many extra absorption peaks, some of which justify the explanation given above for bands marked d in Figure 5.7.

The line intensities in the two spectra do differ, as may be expected if the quantum yield for Si is dependent upon the initially prepared $^1B_1$ rovibronic state. The absorption spectrum is not affected by differences in fluorescence quantum yield, whereas the intensities displayed in the fluorescence excitation spectrum are proportional to the product of absorption line strength and fluorescence quantum yield. Hence, the one-to-one correlation that we observe between the PHOFEX spectrum, and these regions of the absorption spectrum, encouraged McKay et al. to conclude that production of $Si(^3P)$ is associated with the predissociation of rovibronic states of SiH_2 $\tilde{A}\,^1B_1$.

Similar measurements obtained for other regions of the $\tilde{A}\,^1B_1 \leftarrow \tilde{X}\,^1A_1$ SiH_2 band system confirmed their general conclusions, i.e., excitation of short-lived rovibronic states of $\tilde{A}\,^1B_1$ SiH_2 is correlated with strong $Si(^3P)$ PHOFEX signals.

Any $Si(^3P)$ observed in a PHOFEX spectrum from an alternative precursor must also be a result of the three-laser process (IR-MPD–probe–pump), and the precursor must be a primary product of the IR-MPD process. They believe that n-butylsilylene, $CH_3(CH_2)_3SiH$, is the most probable candidate for an alternative silicon source, ethylsilylene. Any n-butylsilylene is then assumed to be produced in the initial IR-MPD process, and will absorb in the SiH_2 $\tilde{A}\,^1B_1 \leftarrow \tilde{X}\,^1A_1$ wavelength region probed in these experiments. They then expect that, upon excitation, the n-butylsilylene will dissociate to form $Si(^3P)$ + butane. The $Si(^3P)$ atoms would result from a predissociative triplet surface coupled to the singlet state that was initially excited. The resultant PHOFEX spectrum, due to $Si(^3P)$ produced by predissociation of excited state n-butylsilylene, would then be indistinguishable from that produced by predissociation of excited state silylene, in the present experimental configuration.

5.1.3 LRAFKS (Laser Resonance Absorption Flash Kinetic Spectroscopy) Apparatus

The experimental methods of Jasinski and Chu [10, 11] are rather different. The Laser Resonance Absorption Flash Kinetic Spectroscopy (LRAFKS) apparatus and the inherently high resolution of the ring laser used provides Doppler-limited spectra. The transition is detected in direct absorption, so that accurate line intensities can be determined without considering quantum state dependences of fluorescent lifetimes or collisional quenching. A scheme of the apparatus is shown in Figure 5.9. The transition of interest, a $15\,cm^{-1}$

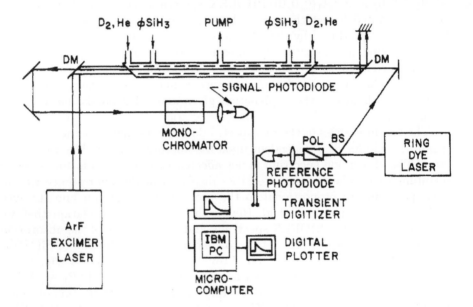

Figure 5.9: Schematic of the LRAFKS apparatus. The dye laser is multipassed three times through the cell by using additional mirrors (not shown). The monochromator serves as a band-pass filter to reject stray excimer laser light. DM are dichroic mirrors, BS is a variable beam splitter, POL is a polarizer used to balance the dye laser intensities incident on the signal and reference photodiodes. The transient digitizer is triggered by a trigger pulse from the excimer laser. Reprinted with permission from Jasinski *J. Phys. Chem.* 86:555, 1986 [10]. Copyright (1988) American Chemical Society.

composite spectrum of overlapped $3\,cm^{-1}$ single frequency scans of the transient absorption produced by 193 nm photolysis of 2 mTorr of phenylsilane in 2 Torr of helium, is shown in Figure 5.10. The spectral region from 17242 to 17352 has been scanned in this manner. This $100\ cm^{-1}$ is around the origin of the $(0, 2, 0) \leftarrow (0, 0, 0)$ vibronic band, and is in the region of high line density. Approximately 100 single rotational lines were observed. Of these 50%

were assigned to known SiH_2 transitions, using the original flash spectroscopic studies of Dubois and his colleagues [1, 2, 4, 5].

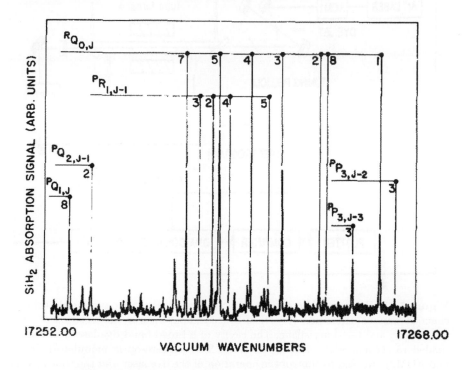

Figure 5.10: Composite spectrum of overlapped 3 cm^{-1} single frequency dye laser scans of a region near the band center of the $^1B_1(0,2,0) \leftarrow {}^1A_1(0,0,0)$ vibronic transition of SiH_2 produced in the 193 nm photolysis of phenylsilane. Line assignments are from Dubois [4]. Reproduced from Jasinski and Chu, *J. Chem. Phys.* 88:1678, 1988 [11], with the permission of AIP Publishing.

5.1.4 Role of Silylene in the Pyrolysis of Silane and Organosilanes

In the work of O'Brien and Atkinson [12] enhanced absorption sensitivity is obtained by placing the species of interest inside the optical cavity of a longitudinally multimode laser (intracavity laser spectroscopy, ILS). Absorptions by intracavity species having spectroscopic transitions within the spectral profile of the ILS laser constitute lass within the resonator. If the spectral widths of the absorption transitions are significantly narrower than the spectral distribution of the laser gain curve, these absorption losses are wavelength selective. This alters the competition that occurs as the modes of the resonance cavity compete for the same gain centres. As a result, the absorption spectrum of the intracavity absorber is superimposed on the output of the ILS laser. This is shown schematically in Figure 5.11.

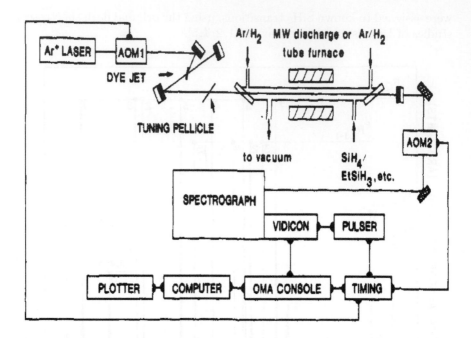

Figure 5.11: ILS spectrometer used to monitor gas-phase SiH_2 produced during the chemical vapour deposition (CVD) of Si-containing films by dissociation of silane, ethylsilane, and tert-butylsilane. The cavity of a broad-band dye laser has been extended to accommodate a CVD chamber. Two acousto-optic modulators (AOM1 and AOM2) are used to control the operation of the dye laser and the time at which its output is examined. AOM1 modulates the intensity of the Ar^+ laser radiation reaching the dye jet above and below the threshold value required for dye laser operation. AOM2 diverts part of the ILS dye laser output beam into a spectrometer after the dye laser has operated for a well-defined period of time, the generation time (I8). The wavelength-dispersed radiation exiting the spectrometer is detected by an intensified vidicon camera. The various silanes are introduced to the CVD chamber via the port labeled $SiH_4/EtSiH_3$, etc., and are decomposed in the intracavity CVD chamber by pyrolysis in a tube furnace or by microwave discharge in an Evenson–Broida cavity. Reprinted with permission from O'Brien and Atkinson, *J. Phys. Chem.* 92:5782, 1988 [12]. Copyright (1988) American Chemical Society.

Figure 5.12 shows ILS spectral profiles at a range of temperatures. In Figure 5.13, the ILS spectrum for the pyrolysis of 31% silane in H_2 at 5.9 Torr of total pressure and $T = 815°C$ ($t_g = 66\,\mu s$) is shown. This spectrum lies in the region of the strongest transitions of the $(0,2,0)' - (0,2,0)''$ of SiH_2. The assignment markers used in this figure are derived from the listing of Dubois [2,4].

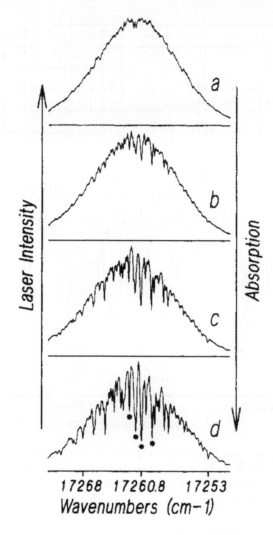

Figure 5.12: ILS spectral profiles for the pyrolytic decomposition of 30% silane in H_2 at 6 Torr of total pressure and with T_h values of (a) 600, (b) 657, (c) 710, and (d) 813 °C ($t_g = 125$ ps for all four temperatures). The horizontal lines correspond to zero laser intensity. Reprinted with permission from O'Brien and Atkinson, *J. Phys. Chem.* 92:5782, 1988 [12]. Copyright (1988) American Chemical Society.

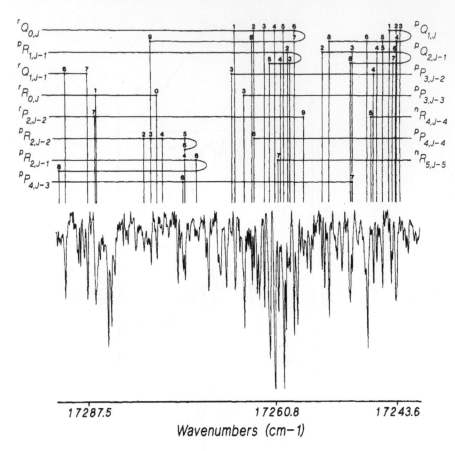

Figure 5.13: ILS spectrum for the pyrolysis of 31% silane in H_2 at 5.9 Torr of total pressure and with $T_h = 815°C$ ($t_g = 66\,\mu s$). Reprinted with permission from O'Brien and Atkinson, *J. Phys. Chem.* 92:5782, 1988 [12]. Copyright (1988) American Chemical Society.

5.1.5 Measurement of SiH$_2$ Densities in an RF-Discharge Silane Plasma Used In the Chemical Vapor Deposition of Hydrogenated Amorphous Silicon Film

Tachibana, Shirafuji, and Matsui investigated the role of SiH$_2$ in the deposition mechanism of hydrogenated amorphous silicon (a-Si:H) films. Silylene radicals were detected by intracavity laser absorption spectroscopy in a parallel plate radio-frequency (RF)-discharge system used in the chemical vapor deposition of hydrogenated amorphous silicon. The experimental setup is shown in Figure 5.14

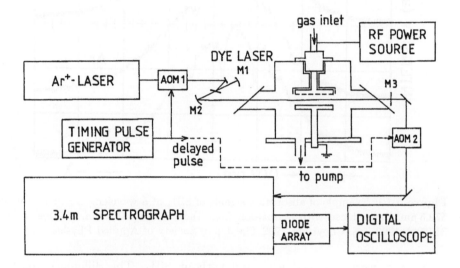

Figure 5.14: Experimental setup of intracavity laser absorption spectroscopy. Reproduced with permission from Tachibana et al., *Jpn. J. Appl. Phys.* 31:2588, 1992 [13]. Copyright 1992 The Japan Society of Applied Physics.

When SiH$_4$ was introduced into the argon plasma, noticeable peaks other than those from Ar and impurities were found, as shown in Figure 5.15.

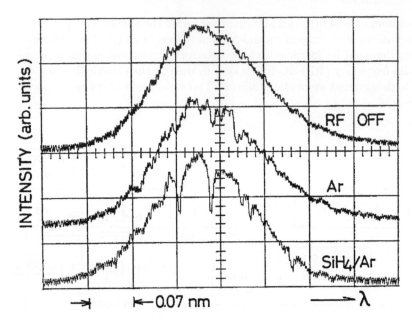

Figure 5.15: Example of absorption signals of SiH$_2$ at a wavelength region around 579.5 nm. Reproduced with permission from Tachibana et al., *Jpn. J. Appl. Phys.* 31:2588, 1992 [13]. Copyright 1992 The Japan Society of Applied Physics.

Spectra were obtained with and without SiH$_4$. The difference spectra were connected together over a range from 17230 cm^{-1} to 17290 cm^{-1} and converted into absorption coefficients. The spectrum obtained in this way is shown in Figure 5.16. Most of the peaks can be assigned to $\tilde{A}\,^1B_1(0,2,0) - \tilde{X}\,^1A_1(0,0,0)$ rovibrational transitions of SiH$_2$.

The absolute concentration of the detected SiH$_2$ radicals could not be measured directly, but was estimated by an indirect comparative procedure using two different calibration methods. Densities ranging from 2×10^8 cm^{-3} to 6×10^9 cm^{-3} were found, depending on the input RF power and on the partial pressure of SiH$_4$. The flux of SiH$_2$ onto the substrate electrode was estimated from the gradient of the spatial distribution. A value of less than 10^{13} cm^{-2}s^{-1} was found, suggesting that the direct contribution of SiH$_2$ radicals to the film deposition is small, though SiH$_2$ is known to play an important role in silicium gas phase reactions.

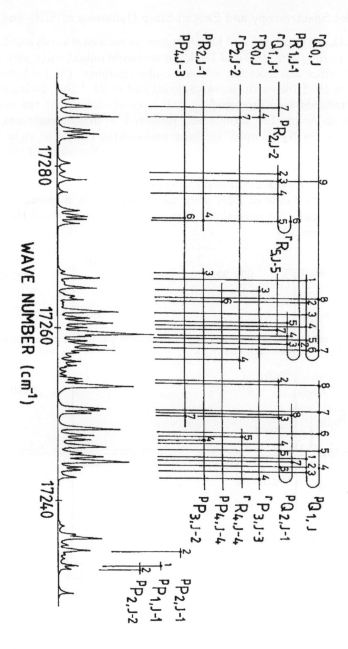

Figure 5.16: Assigned rotational components of the $(0,2,0)' - (0,0,0)''$ band in the $\tilde{A}\,^1B_1 - \tilde{X}\,^1A_1$ transition of SiH_2. Reproduced with permission from Tachibana et al., *Jpn. J. Appl. Phys.* 31:2588, 1992 [13]. Copyright 1992 The Japan Society of Applied Physics.

5.1.6 Jet Spectroscopy and Excited State Dynamics of SiH_2 and SiD_2

In the work of Fukushima and his colleagues on excited state dynamics [14,15], reported in the years 1992 and 1994, a homemade pulsed valve with a $500\,\mu m$ diameter orifice was mounted onto a vacuum chamber. The background pressure was $\approx 10^{-6}$ Torr with the valve closed and $\approx 10^{-4}$ Torr during operation. Silylene radicals were produced by ArF laser photolysis, at the nozzle exit, seeded in an Ar jet. LIF signals were observed 35 mm downstream from the orifice, using an N_2 pumped dye laser intersecting the jet at right angles to the direction of propagation of the gas pulse. The linewidth of the dye laser was $\approx 0.3\,cm^{-1}$.

$\tilde{A}^1 B_1 - \tilde{X}^1 A_1$ LIF excitation spectra of eight vibronic bands of SiH_2 and eleven vibronic bands of SiD_2 were measured in the wavelength region of 460–640 nm under jet conditions. The spectra of five of the SiH_2 bands are shown in Figure 5.17.

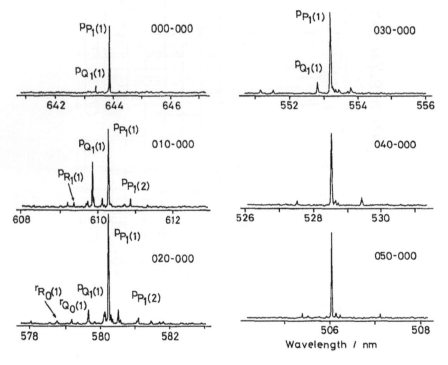

Figure 5.17: LIF excitation spectra of the SiH_2 $\tilde{A}^1 B_1(0, v_2', 0) - \tilde{X}^1 A_1(0, 0, 0)$ $v_2' = 0 - 5$ transition. Signals are integrated for $0\text{-}5\,\mu s$. Reproduced from Fukushima et al., *J. Chem. Phys.* 96:44, 1992 [14], with the permission of AIP Publishing.

Vibrational assignments follow those of Dubois [1, 2, 5] at room temperature. Lines corresponding to the r subband were almost absent in their excitation spectra, but were present in the absorption spectrum of Dubois.

The lifetimes of vibronic levels in the $\tilde{A}\,{}^1B_1$ state were determined by time-dependent fluorescence spectroscopy. These levels were $v_2' = 0 - 7$ for SiH_2 and $v_2' = 0 - 10$ for SiD_2. Examples of these signals are shown in Figure 5.18. A shorter than expected fluorescence lifetime was found for the $v_2' = 7$ state of SiH_2, which indicates the opening of a non-radiative decay channel. The experiments also permitted an estimation of the electronic transition moment.

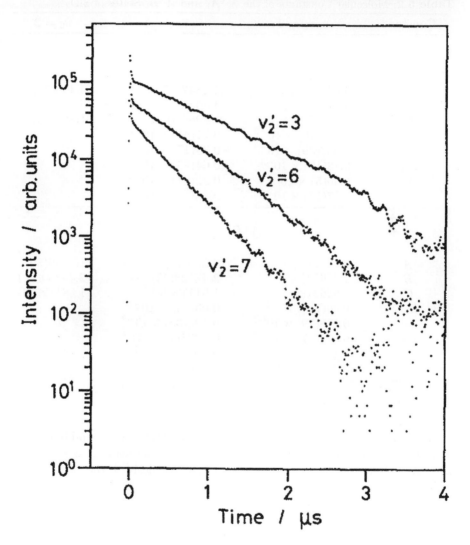

Figure 5.18: Time resolved profiles of fluorescence intensity for SiH_2 excited at ${}^PP_1(1)$, $v_2' = 3, 6$ and 7. (The weak oscillation observed on the weak signal region after $3\,\mu s$ is an apparatus artefact.) Reproduced from Fukushima et al., *J. Chem. Phys.* 96:44, 1992 [14], with the permission of AIP Publishing.

In their 1994 paper, Fukushima and Obi [15] continued their work on deuterated silylene, SiD_2, with the high-resolution LIF excitation apparatus. The rotational structure of five vibronic bands was analysed and led to the rotational parameters presented in Table 5.2.

Table 5.2: Molecular Constants of the $\tilde{X}\,^1A_1$ and $\tilde{A}\,^1B_1$ States of SiD_2

Constant	$v_2 = 0$	$v_2 = 1$	$v_2 = 2$
		$\tilde{X}\,^1A_1$ state	
A	4.3345(11)	4.4343(17)	4.435(78)
B	3.51856(53)	3.5631(13)	3.5740(37)
C	1.91945(25)	1.89854(60)	1.8837(31)
Δ_J	$0.102(10) \times 10^{-3}$	$0.71(41) \times 10^{-4}$	
Δ_{JK}	$-0.382(39) \times 10^{-3}$	$-0.38(17) \times 10^{-3}$	
Δ_K	$0.83(10) \times 10^{-3}$	$0.64(25) \times 10^{-3}$	
δ_J	$0.46(5) \times 10^{-4}$	$0.22(21) \times 10^{-4}$	
δ_K	$-0.37(17) \times 10^{-4}$	$0.48(61) \times 10^{-4}$	
		$\tilde{A}\,^1B_1$ state	
A	9.62878(93)	10.58680(96)	11.7859(30)
B	2.45623(72)	2.47128(42)	2.49230(71)
C	1.92621(84)	1.91119(43)	1.89112(53)
Δ_J	$0.06(7) \times 10^{-4}$	$0.033(4) \times 10^{-4}$	
Δ_{JK}	$-0.178(79) \times 10^{-3}$	$-0.425(33) \times 10^{-3}$	
Δ_K	$0.812(17) \times 10^{-2}$	$0.1238(19) \times 10^{-1}$	
δ_J	$0.15(3) \times 10^{-4}$	$0.10(1) \times 10^{-4}$	
δ_K	$-0.29(35) \times 10^{-3}$	$0.23(15) \times 10^{-3}$	

Data are given in cm^{-1}. Values in parentheses denote one standard deviation and apply to the last digits of the constants.

Intensity anomalies were found in the rotational structures and interpreted in terms of a non-radiative predissociation mechanism leading to $Si\,(^3P)$ and $D_2\,(^1\Sigma_g^+)$. Fluorescence quantum yields were then derived, they are reported in Table 5.3.

Table 5.3: Estimated Fluorescence Quantum Yields In K_a Stacks of the $\tilde{A}\,^1B_1$ State of SiD_2

K_a	symmetry	$v_2' = 2$		$v_2' = 1$	
		k_{nr}/k_f	ϕ_f	k_{nr}/k_f	ϕ_f
0	a	0	1	0	1
	s	0	1	0	1
1	a	20	0.047	1	0.50
	s	40	0.024	2	0.33
2	a	80	0.012	4	0.20
	s	160	0.0062	8	0.11
3	a	180	0.0055	9	0.10
	s	360	0.0028	18	0.052
4	a	$\cdots$	0	16	0.059
	s	$\cdots$	0	32	0.030

k_{nr}/k_f is the ratio of non-radiative and fluorescence rate constants, ϕ_f is the fluorescence quantum yield. The ratio for symmetric states (s) is twice that for antisymmetric states (a), reflecting the different density of states of the dissociation products. s-states lead to *ortho* D_2, and a-states to *para* D_2, respectively.

5.1.7   Intracavity Laser Absorption Spectroscopy, ICLAS

At the end of the 1990s, interest in silylene became much greater because of the importance of mechanistic chemistry of silicon hydride molecules in the production of silico-radical by using the intracavity laser absorption films. The $\tilde{A}\,^1B_1(0,0,0,) - \tilde{X}\,^1A_1(0,0,0)$ band of silylene was studied in detail by Escribano and Campargue in 1998 [3]. They generated the spectrum of the silylene radical by using the intracavity laser absorption spectroscopy technique (ICLAS). The silylene radical was generated by a continuous discharge (40 mA, 1 kV) in a slowly flowing mixture of silane (5%) in argon. The total pressure was about 1 Torr and gases were pumped out with a mechanical pump. The plasma tube (80 cm length, 3.6 cm diameter) was fitted with Brewster angle windows. The flow of argon entered the cell close to the centre of the cell. This arrangement minimizes the deposition of silican films on the Brewster windows, as observed on the walls of the cell. The distance between the two cylindrical electrodes was 60 cm.

The ICLAS technique [3] is based on the high sensitivity of a broadband laser to intracavity losses such as absorption. The absorption lines appear superimposed on the broadband spectra of the dye laser (DCM) pumped by an argon laser. The spectrum is dispersed by a high resolution grating spectrograph with a resolving power up to 800,000. The spectra are recorded by a 1024 photodiode array. The dye laser is time resolved by using two synchronized acousto-optic modulators. The generation time, t_g, between the start of the laser and the recording of the spectrum gives, directly, the equivalent path lengths, $l_{eq} = (l/L)ct_g$. Here, (l/L) is the ratio of the length of the cell to the optical length of the dye laser cavity.

In their present experiments, they inserted a plasma tube into the dye laser cavity. The generation times used ranged from 20 to 70 μs (absorption equivalent path lengths between 3.7 and 13.0 km). The occupation ratio of the cavity by the cell was 62%. In order to reduce the number of spectra needed to cover the whole band, 750 cm^{-1}, the spectrograph was run with a half width at half maximum of about 0.05 cm^{-1}. The estimated wavenumber accuracy is about 0.01 cm^{-1}. The intensity of the $\tilde{A}\,^1B_1(0,0,0) \leftarrow \tilde{X}\,^1A_1(0,0,0)$ band is 22% of the $\tilde{A}\,^1B_1(0,2,0) \leftarrow \tilde{X}\,^1A_1(0,0,0)$ transition, which was also studied by Campargue, Romanini, and Sadeghi in 1998 [16].

The ICLAS spectrum in the range between 15350 cm^{-1} and 15800 cm^{-1} is presented in Figure 5.19. Assignment of this spectrum started from the central region. The data were then fitted to determine the rovibrational parameters in Table 5.4. Figure 5.20 shows the central part of the spectrum together with its simulation.

Table 5.4: The rovibrational parameters of the $(v_1, v_2, v_3) = (0,0,0)$ level of the $\tilde{X}\,^1A_1$ and $\tilde{A}\,^1B_1$ electronic states of SiH$_2$

Parameter	$\tilde{X}\,^1A_1$	$\tilde{A}\,^1B_1$
ν_0	0	15547.7730(93)
A	8.09617(25)	18.3241(21)
B	7.02262(25)	4.89951(87)
C	3.70058(21)	3.76611(31)
δ_J	$0.20238(89)\times10^{-3}$	$0.412(28)\times10^{-4}$
δ_K	0	0
Δ_J	$0.4341(22)\times10^{-3}$	$0.895(53)\times10^{-4}$
Δ_{JK}	$-0.14013(89)\times10^{-2}$	$-0.337(63)\times10^{-3}$
Δ_K	$0.22995(86)\times10^{-2}$	$0.3516(14)\times10^{-1}$
ϕ_J	0	0
ϕ_{JK}	0	$-0.399(70)\times10^{-5}$
ϕ_{KJ}	0	$0.235(12)\times10^{-4}$
ϕ_K	0	$0.1843(28)\times10^{-3}$
L_K	0	$-0.816(19)\times10^{-6}$
rms	0.045	0.038

The ground state parameters were fitted from the data of Yamada et al. [17], Dubois [2], and the present work [3]. Parameters without uncertainties were constrained to 0. All values in cm^{-1}. The total number of data included in the fit is 1315 corresponding to 54 transitions of Reference [17] affected with weight 10, 988 data of Reference [2] with a weight 0.1, and 273 transitions of the present work [3] with weight 1. With these data, 851 ground state combination differences were formed.

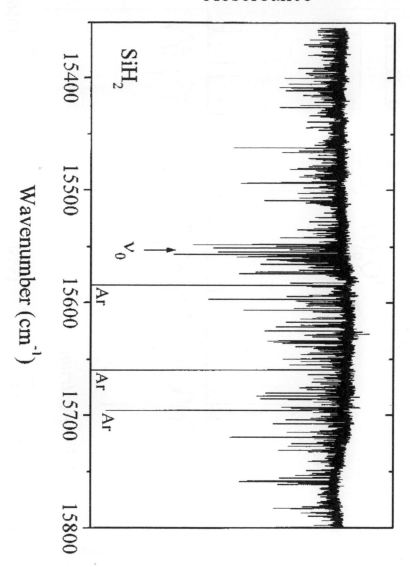

Figure 5.19: The ICLAS spectrum of SiH_2 generated in a continuous discharge, (40 mA, 1 kV), in a mixture of silane (5%) in argon. The total pressure was about 1 Torr, and the equivalent path length was about 11 km ($t_g = 60$ μs). The band origin ν_0 of the $\tilde{A}\,^1B_1(0,0,0) - \tilde{X}\,^1A_1(0,0,0)$ transition is indicated as well as some absorption lines from excited states of Ar. Reproduced from Escribano and Campargue, *J. Chem. Phys.*, 108:6249, 1998 [3], with the permission of AIP Publishing.

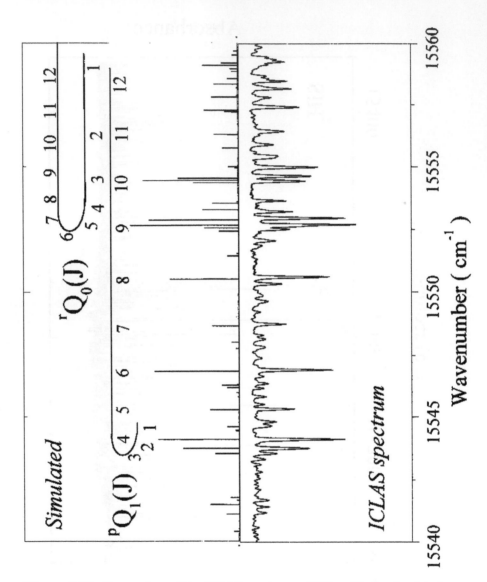

Figure 5.20: Comparison of the ICLAS spectrum of SiH$_2$ with the simulated spectrum obtained using the rotational constants of Table 5.4. Reproduced from Escribano and Campargue, *J. Chem. Phys.*, 108:6249, 1998 [3], with the permission of AIP Publishing.

Analysis of the upper-state levels revealed local perturbations, that are due to spin-orbit coupling between the $^1B_1$ and the $^3B_1$ states [18]. The perturbed levels are collected in Table 5.5.

Table 5.5: Energy (In cm^{-1}) of the Perturbed $J'K_a'K_c'$ Rotational Levels of the $\tilde{A}\,^1B_1(0,0,0)$ State and Energy Shift Δ (in 10^{-2}cm^{-1}) Due to the Perturbation

J'	K_a'	K_c'	E_v'	Δ	N^a	J'	K_a'	K_c'	E_v'	Δ	N^a
3	1	3	15610.33	9	2	4	1	4	15642.34	-15	3
5	1	5	15682.75	-6	3	3	2	1	15655.63	15	4
4	2	2	15691.13	37	5	7	2	5	15851.58	-53	3
8	2	6	15924.12	-59	3	3	2	2	15655.74	60	3
5	2	4	15732.75	-12	4	8	2	7	15912.90	14	2
4	3	1	15914.98	14	4	6	3	3	15853.22	-59	5
8	3	5	15984.83	-52	2	4	3	2	15758.19	38	4
5	3	3	15801.02	-31	4	7	3	5	15914.37	-16	5
6	4	2	15946.46	53	5	7	4	3	16005.38	50	2
6	4	3	15946.05	13	2	8	4	3	16076.27	-23	3
5	5	0	16008.55	-12	2	6	5	1	16061.19	20	3
8	5	3	16192.36	68	4	5	5	1	16008.53	-14	3
6	5	2	16061.23	26	2	7	5	3	16122.53	16	4

^{a}Number of observed transitions used to calculate the upper-state energy level.

In order to understand the variations of the absorption and emission spectra of SiH$_2$, some comparisons were made between the ICLAS spectrum of the central region of SiH$_2$ $\tilde{A}\,^1B_1(0,0,0) \leftarrow \tilde{X}\,^1A_1(0,0,0)$ and the equivalent spectra of Dubois, $\tilde{A}\,^1B_1(0,v_2',0) \leftarrow \tilde{X}\,^1A_1(0,0,0)$, where $v_2' = 1, 2$, and 3. The behaviour of the central $^rQ_0(J)$ and the $^pQ_1(J)$ branches are shown in Table 5.6.

Table 5.6: A Comparison of the Rotational Assignments and Wavenumbers For Four Rovibronic Transitions of SiH_2

$\tilde{A}^1B_1$	$\tilde{X}^1A_1$	$\tilde{A}^1B_1$ (0,0,0)	$\tilde{A}^1B_1$ (0,1,0)	$\tilde{A}^1B_1$ (0,2,0)	$\tilde{A}^1B_1$ (0,3,0)
1_{01}	1_{11}	15544.592	16402.06	17244.56	18093.59
2_{02}	2_{12}	15543.699	16400.21	17243.61	18097.75
3_{03}	3_{13}	15543.495	16397.79	17242.92	18096.65
4_{04}	4_{14}	15544.068	16396.27	17243.55	18093.34
5_{05}	5_{15}	15545.264		17245.52	18090.8
6_{06}	6_{16}	15546.852	16397.20	17247.65	18092.90
7_{07}	7_{17}	15548.632			
8_{08}	8_{18}	15550.518	16398.4	17253.01	
9_{09}	9_{19}	15552.408			
$10_{0\,10}$	$10_{1\,10}$	15554.322	16402.06		
1_{11}	1_{01}	15559.105	16420.34	17266.47	18121.02
2_{12}	2_{02}	15556.302	16414.95	17263.77	18120.07
3_{13}	3_{03}	15554.526	16411.24	17262.26	18115.42
4_{14}	4_{04}	15553.079	16407.89	17260.8	18107.92
5_{15}	5_{05}	15552.570	16405.62	17259.5	18101.39
6_{16}	6_{06}	15552.570	16403.88	17258.05	16101.26
7_{17}	7_{07}	15552.835	16402.06	17258.05	
8_{18}	8_{08}	15553.527	16401.1	17264.11	
9_{19}	9_{09}	15554.526	16400.63	17278.45	
$10_{1\,10}$	$10_{0\,10}$	15555.791	16403.88		

Rovibronic transitions $\tilde{A}^1B_1(0,0,0) \leftarrow \tilde{X}^1A_1(0,0,0)$ were recorded by Escribano and Campargue [3], those of $\tilde{A}^1B_1(0,v_2',0) \leftarrow \tilde{X}^1A_1(0,0,0)$, $v_2' = 1, 2$, and 3 by Dubois [2]. Note the decrease in the number of transitions $v_2' = 1, 2$, and 3 as v_2' increases.

5.1.8 Measurement of SiH_2 Density in a Discharge by Intracavity Laser Absorption Spectroscopy and CW Cavity Ring-Down Spectroscopy

Cavity ringdown spectroscopy is a very sensitive method to measure absorption cross sections. The first applications employed pulsed lasers, however the spectral resolution can be improved considerably by the use of continuous wave (CW) lasers, with the frequency tuned to one of the cavity resonances. Campargue, Romanini, and Sadeghi [16] applied this new method to the detection of SiH_2 in an argon-5% silane DC discharge. Their experimental setup is shown in Figure 5.21.

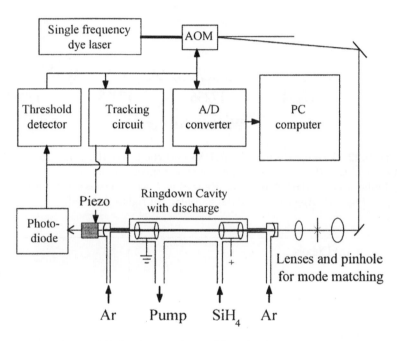

Figure 5.21: A simplified scheme of the CW-CRDS setup for the concentration measurements of SiH_2 in an Ar–5% SiH_4 discharge. © IOP Publishing. Reproduced with permission from Campargue et al., *J. Phys. D: Appl. Phys.*, 31:1168, 1998. All rights reserved.

The performances of CW-CRDS and ICLAS are compared in Figure 5.22. The spectral resolution is slightly better for the CRDS spectrum because the resolution of the ICLAS setup is limited by the apparatus function of the spectrograph, but the signal-to-noise ratios are comparable. The sensitivity and resolution obtained here are much better than those in the two former studies, by O'Brian and Atkinson [19] and Jasinski and Chu [11], in which SiH_2 was observed by ICLAS in Ar-SiH_4 discharges. A comparison of the ICLAS spectrum with the original stick spectrum by Dubois [4] is presented in Figure 5.23.

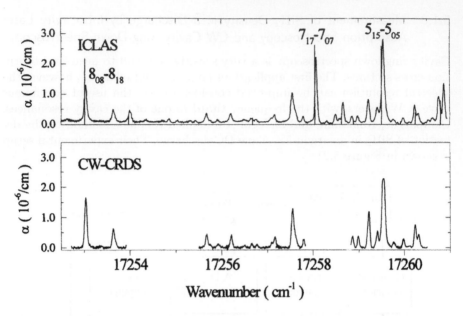

Figure 5.22: A comparison of the ICLAS and the CW-CRDS spectra of SiH$_2$ near the centre of the (0,2,0)–(0,0,0) band. The stick spectrum on the ICLAS spectrum is from Duboisi [4]. © IOP Publishing. Reproduced with permission from Campargue et al., *J. Phys. D: Appl. Phys.*, 31:1168, 1998. All rights reserved.

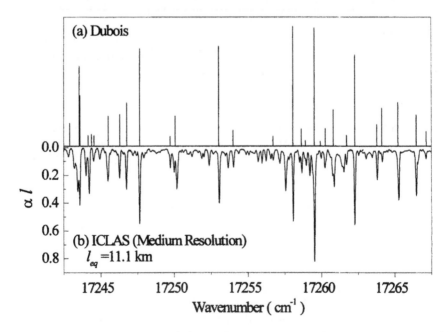

Figure 5.23: A comparison of the stick spectrum from Dubois with the ICLAS spectrum. © IOP Publishing. Reproduced with permission from Campargue et al., *J. Phys. D: Appl. Phys.*, 31:1168, 1998. All rights reserved.

5.1.9 Renner–Teller and Spin-Orbit Interactions in SiH_2

The initial study of the $\tilde{A}\,^1B_1 - \tilde{X}\,^1A_1$ transition was carried out by Dubois [2] in 1968. He observed the absorption spectrum of the transient produced by flash discharges in a variety of precursors. From the rotational analysis of this spectrum, he obtained the rotational term values of the (0,0,0) vibrational level of the ground state and the (0,1,0), (0,2,0) and (0,3,0) levels of the excited state. Subsequently, in 1975, Dubois, Duxbury and Dixon [5] extended the excited state levels to (0,4,0), (0,5,0) and (0,6,0). They showed that the two combining states, the $\tilde{X}\,^1A_1$ and the $\tilde{A}\,^1B_1$ states, are derived from a $^1\Delta$ electronic state split by static Renner interaction. The vibronic pattern in the excited state was shown to be due to both large amplitude vibration and the Renner effect. The barrier to linearity is ca. $8000\,cm^{-1}$. The (0,0,0) levels of the excited state were measured subsequently by Escribano and Campargue [3] in 1998.

In 1993, Duxbury, Alijah, and Trieling [18] analysed these perturbations. They discussed the origins of the perturbations observed in the spectrum of SiH_2, both level shifts and lifetimes, and gave a semiquantitative explanation of their size and distribution by applying the model of Alijah and Duxbury [6], developed originally for CH_2, to predict the pattern of coupling between the states. Singlet-triplet interactions in the excited $^1B_1$ state arise due to second-order coupling mediated by the lowest singlet state, $^1A_1$: There is direct Renner–Teller coupling between the $^1A_1$ and $^1B_1$ states and singlet-triplet coupling between the $^1A_1$ and $^3B_1$ states.

Perturbations of the rotational levels in the excited $^1B_1$ state appear even for $K_a = 0$, where no Renner–Teller effect is present. It is instructive to compare the rotational structure in the upper state of SiH_2 with that of the equivalent state of PH_2. Even though PH_2 is a doublet system and necessitates consideration also of spin-rotation interaction, its rotational states can be fitted to a standard Hamiltonian with consistent accuracy for the lowest vibrational states [20], whereas this is not the case for SiH_2. As can be seen from Figures 5.24 and 5.25, respectively, in which the effective rotational constants, $\bar{B}_{eff} = E_{rot}/J(J+1)$, are plotted as a function of $J(J+1)$ for the two molecules, the $(0,3,0)$ band of SiH_2 is heavily perturbed. In the absence of Renner–Teller coupling, these perturbations are interpreted as due to Fermi resonances.

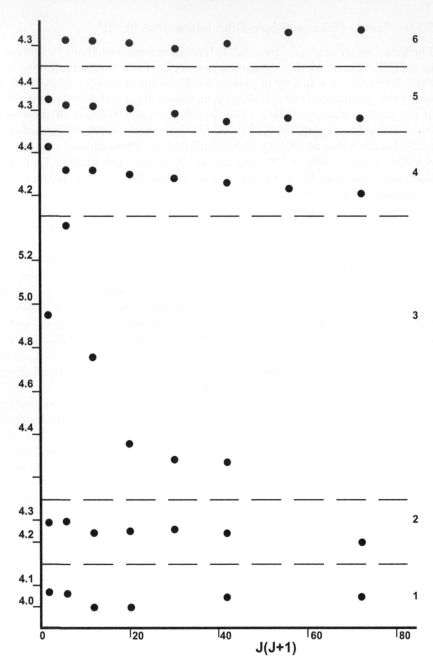

Figure 5.24: Observed values of $\bar{B}_{eff}$ derived from the $K = 0$ term values of the $^1B_1$ state of SiH_2. Reproduced from Duxbury et al., *J. Chem. Phys.* 98:811, 1993 [18], with the permission of AIP Publishing.

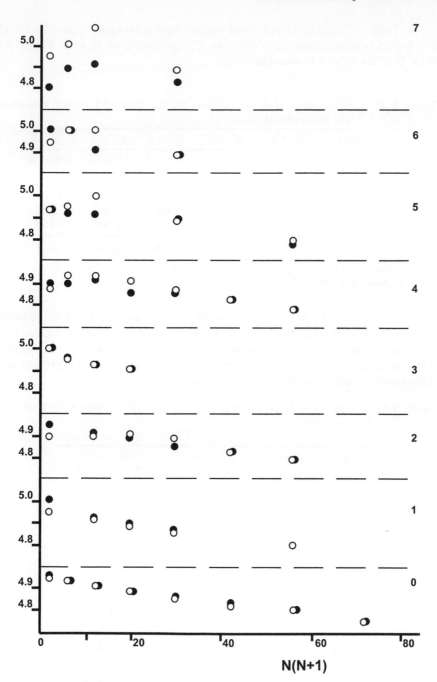

Figure 5.25: Observed values of $\bar{B}_{eff}$ derived from the $K = 0$ term values of the $^2A_1$ state of PH_2. Open circles: $J = N - \frac{1}{2}$. Full circles: $J = N + \frac{1}{2}$. Despite the doublet splitting in PH_2, the SiH_2 term values show the effects of much larger local perturbations. Reproduced from Duxbury et al., *J. Chem. Phys.* 98:811, 1993 [18], with the permission of AIP Publishing.

In Table 5.7, the observed band origins and rotational constants for the five lowest bending states are collected. The deviation of $\bar{B}_{eff}$ in Figure 5.24 from $\bar{B}_{obs}$ for $v_2' = 3$ is remarkable.

Table 5.7: Derived Values of ν, A_v, B_v and C_v for Excited Bending Vibrational States of the $\tilde{A}^1B_1$ State of SiH$_2$

Constants	$v_2' = 0$	$v_2' = 1$	$v_2' = 2$	$v_2' = 3$	$v_2' = 4$
ν_{obs}	15547.7730 (a)	16405.7 (b)	17247.8 (b)	18095.5 (b)	18935.8 (b)
A_{obs}	18.3241	20.59	24.80	29.83	39.07
B_{obs}	4.89951	4.951	4.845	5.967	5.163
C_{obs}	3.76611	3.639	3.649	3.307	3.753
$\bar{B}_{obs}$	4.33281	4.295	4.247	4.637	4.458

Data are in cm^{-1}. (a) From Escribano and Campargue [3]; (b) From Dubois [2,5]. $\bar{B} = (B+C)/2$.

Table 5.7 also contains the data for the vibrational ground state measured by Escribano and Campargue [3] in 1998, five years after the original analysis. As a consistency check, we have evaluated, in Table 5.8, combination differences using their data and that obtained by Dubois. The ground state differences are identical to within error limits.

Table 5.8: The $\tilde{A}^1B_1 - \tilde{X}^1A_1$ Absorption Spectrum of the Free SiH$_2$ Radical

$J'K_a'K_c'$	$J''K_a''K_c''$	ΔE (cm^{-1})	Diff (cm^{-1})
		$(0,0,0) - (0,0,0)$	
110	000	15570.943	45.558
110	220	15525.385	
211	101	15578.675	73.011
211	321	15505.664	
313	202	15587.355	103.179
313	422	15484.176	
413	303	15597.97	134.092
413	523	15463.878	
514	404	15610.324	164.430
514	624	15445.894	
615	505	15624.143	194.415
615	725	15429.728	

Table 5.8: The $\tilde{A}\,^1B_1 - \tilde{X}\,^1A_1$ Absorption Spectrum of the Free SiH$_2$ Radical

$J'K_a'K_c'$	$J''K_a''K_c''$	ΔE (cm^{-1})	Diff (cm^{-1})
		$(0,1,0) - (0,0,0)$	
110	000	16431.46	45.540
110	220	16385.92	
211	101	16438.53	73.080
211	321	16365.45	
312	202	16446.75	103.220
312	422	16343.53	
413	303	16456.33	134.080
413	523	16322.25	
514	404	16467.79	164.490
514	624	16303.3	
615	505	16479.62	194.340
615	725	16285.28	
		$(0,2,0) - (0,0,0)$	
110	000	17277.69	45.490
110	220	17232.2	
211	101	17286.36	73.060
211	321	17213.3	
312	202	17295.25	103.150
312	422	17192.1	
413	303	17306.36	134.160
413	523	17172.2	
514	404	17320.46	164.430
514	624	17156.03	
615	505	17333.52	194.440
615	725	17139.08	
		$(0,3,0) - (0,0,0)$	
110	000	18131.72	45.550
110	220	18086.17	
211	101	18143.15	73.000
211	321	18070.15	
312	202	18155.17	103.150
312	422	18052.02	
413	303	18166.95	134.040
413	523	18032.91	
514	404	18182.26	164.580
514	624	18017.68	

Table 5.8: The $\tilde{A}\,{}^1B_1 - \tilde{X}\,{}^1A_1$ Absorption Spectrum of the Free SiH_2 Radical

$J'K_a'K_c'$	$J''K_a''K_c''$	ΔE (cm^{-1})	Diff (cm^{-1})
615	505	18202.78	194.440
615	725	18008.34	
	Average $\pm$ Standard Deviation		
110	000		45.535$\pm$0.031
110	220		
211	101		73.038$\pm$0.038
211	321		
312	202		103.175$\pm$0.033
312	422		
413	303		134.093$\pm$0.050
413	523		
514	404		164.483$\pm$0.071
514	624		
615	505		194.409$\pm$0.047
615	725		

The difference spectrum of the $(0,0,0) - (0,0,0)$ absorption spectrum was recorded by Escribano and Campargue [3], those of the $(0,1,0) - (0,0,0)$, $(0,2,0) - (0,0,0)$, and the $(0,3,0) - (0,0,0)$ bands by Dubois [2].

Duxbury, Alijah, and Trieling then reanalyzed some of the original data on the $(0,2,0) - (0,0,0)$ band recorded by Dubois, supplemented by the high resolution spectra of Jasinski, to see whether they could find evidence for the existence of perturbations similar to those observed by Petek et al. [21, 22] for CH_2 using magnetic rotation spectroscopy. In order to refit these data, they used the new ground state rotational constants reported by Yamada et al. [17]. The values for the upper-state rotational constants were now closer to those of the $(0,1,0)$ bending state, which led to the corrections of some of the former assignments shown in Table 5.9.

Table 5.9: Assignments of Perturbation Allowed Transitions With $K_a'' = 1$

J	Dubois	New Assignment
	$^R R_{0,J}$ branch	
2	17295.24	17294.74
3	17306.36	17304.83
4	17320.46	17319.01
	$^P P_{2,J-2}$ branch	
4	17192.11	17191.68
5	17172.27	17170.75
6	17156.03	17154.45
	$^R Q_{0,J}$ branch	
3	17262.26	17260.32[a]
4	17266.80	–
5	17259.50	17259.20[a]
	$^P P_{2,J-1}$ branch	
4	17202.20	17200.22
5	17185.80	–
6	17169.60	17169.12

Data are in cm^{-1}. Assignments of transitions in the
(0,2,0) - (0,0,0) band compared with previous
assignments made by Dubois [2]. All measurements are
from unpublished data of Dubois [4] except for the
absorption spectra of Jasinski.
[a]Line from unpublished data of Jasinski

Next, Duxbury and coworkers analysed the new experimental results for
the fluorescence liftimes of individual rovibronic levels of the $^1B_1$ state that
had been obtained by Thoman and Steinfeld [7], Thoman et al. [8], Francisco et
al. [23] and Fukushima and colleagues [14]. These measurements showed that
there were three regimes with qualitatively different behavior. With low values
of v_2', $v_2' = 0, 1$, and 2, a complex variation of the lifetimes with rovibronic level
structure was observed. A large number of levels have lifetimes of ~ 10 ns,
although there are a significant number with very long lifetimes of up to at
least 1.28 ps. As examples of this, lifetimes of rovibronic levels of the (0,2,0)
vibrational state of $\tilde{A}\,^1B_1$ are given in Table 5.10.

Table 5.10: Observed Rovibronic Lifetimes of Levels of the (0,2,0) Vibrational State of the $\tilde{A}\,^1B_1$ Electronic State of SiH_2

JK_aK_c	τ(ns)	JK_aK_c	τ(ns)	JK_aK_c	τ(ns)
000	898				
101		111	17±1	110	1054.20±7
202	151	212	11±4	211	137.8
303	16±3	313	150.12	312	9
404	12	414	13±4	413	86.9±4
505	7	515	109.8	514	10
606	8±2	616	60	615	77.5
707	106.28	717	14±7	716	
808	30	818		817	817±80.11
221		220	15±8		
322	L^a, 10±6	312	122.11		
423	40.9	422	5		
524	678.8±2	523			
625	8	624			
726	8±2	725			
827		826	109.20		
928	8±2	927			
331	7	330	8		
432	94.10	431	5±1		
533		532			
634	7	633	1082.16±7		
836	19	835			
937	10	836			

The majority of the lifetimes measured are less than 50 ns (after Thoman et al. [8]). Note the very long lifetime of the rotationless level 0_{00}.
$^a L$ is a long lifetime (McKay et al. [9]).

In this table, the lifetimes have been ordered by K_a, and J. It can be seen that although the measured lifetime of the 0_{00} level is $\sim 1\ \mu s$, most of the rovibronic lifetimes measured by Thoman et al. lie between 5 ns and 20 ns and exhibit no obvious J or K_a dependence.

Figure 5.26 shows a comparison of laser-induced fluorescence and high-resolution absorption spectra. The traces are: (A) High-resolution dye laser absorption spectrum due to Jasinski [10]. The $1_{11} - 2_{21}$, $0_{00} - 1_{10}$ and $1_{10} - 2_{20}$ are taken from (B), since the dye laser spectrum finishes at the $^P Q_{1J}$ branch. (B) Flash discharge absorption spectrum due to Dubois [2, 4]. (C) Laser-induced fluorescence spectrum in the $0 - 50$ ns time window [20]. (D) Laser-induced fluorescence spectrum in the $60 - 260$ ns time window [20]. The short-lived component of the LIF (C) is very similar to the two high-resolution spectra (A) and (B); dashed lines are used to show the corresponding features. Note the very long lifetime of the $0_{00} - 1_{10}$ transition, which involves the rotationless excited state level. The other long-lived transitions seen in (D) do not show any obvious connections with low J transitions. One particular transition marked $\times$ is very strong in (C), of medium intensity in (D), and corresponds to a weak unassigned line in the absorption spectrum.

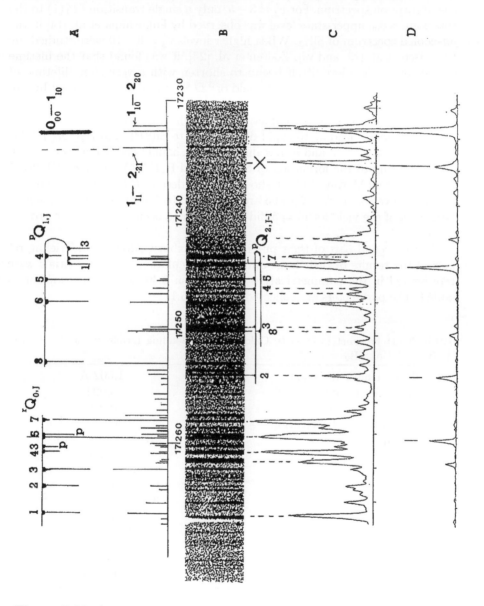

Figure 5.26: A comparison of laser-induced fluorescence and high-resolution absorption spectra near the centre of the $(0,2,0) - (0,0,0)$ band of the $\tilde{A}\,^1B_1 - \tilde{X}\,^1A_1$ electronic transition of SiH_2. See text for further details. Reproduced from Duxbury et al., *J. Chem. Phys.* 98:811, 1993 [18], with the permission of AIP Publishing.

The spectrum of the short lifetime component of the fluorescence correlates well with the absorption spectrum of the $(0,2,0)-(0,0,0)$ band. However, in general, the long lifetime component shown in the figure has little similarity to the absorption spectrum. For $v_2' = 3-5$, only a single transition $^PP_1(1)$ to the rotationless 0_{00} upper-state level was observed by Fukushima et al. [14] in the jet-cooled spectrum of SiH$_2$. When higher levels $v_2' = 6-10$ were studied, by Francisco et al. [23] and van Zoeren et al. [24], it was found that the lifetime of the rotationless level itself began to shorten with a very short lifetime of less than 10 ns for $v_2' > 8$. A high yield of ^1D Si atoms was also found by van Zoeren et al. [24] in PHOFEX experiments. This behavior can be interpreted in terms of a coupling to the triplet manifold which goes through a maximum for $v_2' = 4$ and 5, leading to rapid dissociation to Si(^3P), and H$_2$($^1\Sigma_g^+$). This then closes at $v_2' = 6$, and a new dissociating channel opens between $v_2' = 6$ and 7, leading to the formation of Si (^1D) and H$_2$($^1\Sigma_g^+$). Further PHOFEX experiments by McKay et al. [9] showed that indeed Si (^3P) could be detected when the levels with $v_2' = 2$ and 5 are populated by laser pumping, although a large part of the yield for $v_2' = 5$ did not appear to occur via the same process as in $v_2' = 2$.

In the second part of their publication, Duxbury, Alijah, and Trieling reported the results of their large-amplitude calculations. The potentials were represented in the parametrization of Equation 3.40, with correction terms added. The parameters are collected in Table 5.11.

Table 5.11: Parameters Used to Calculate the Bending Levels of the $\tilde{X}\,^1A_1$ and $\tilde{A}\,^1B_1$ States of SiH$_2$

Bond length variation	$r(\rho = 0)$	1.502 Å
d coefficient of $\tan^2(\rho/2)$	$\tilde{X}$ state	0.024
	$\tilde{A}$ state	0.070
f_{33} for inertial correction		150000.0 cm^{-1}
$\tilde{A}\,^1B_1$ state	f	23229.7 cm^{-1} rad^{-2}
	ρ_m	56.996°
	H_0	7028.2 cm^{-1}
	c_H	-0.106
$\tilde{X}\,^1A_1$ state	f	33487.6 cm^{-1} rad^{-2}
	ρ_m	87.895°
	H_0	22625.0 cm^{-1}
	c_H	-0.04712
	c_1	0.268 cm^{-1}
	c_2	649.7 cm^{-1}

The calculated rovibronic term values reproduce correctly most of the experimentally observed pattern for both SiH_2 and SiD_2. As an example, we present, in Table 5.12, the observed and calculated term values for the $\tilde{A}\,^1B_1$ state up to $J' = 3$.

Table 5.12: Observed and Calculated Rovibronic Levels of the $\tilde{A}\,^1B_1$ State of SiH_2 (T_0/cm^{-1})

JK_aK_c	$v' = 0$	$v' = 1$	$v' = 2$	$v' = 3$	$v' = 4$	$v' = 5$	$v' = 6$	
000	15547.7730	16404.3	17248.2	18096.9	18934.9	19776.0	20618.8	o.
		16401.4	17247.4	18092.4	18936.1	19778.5	20621.2	c.
101		16412.4	17256.8	18106.8	18943.8	19784.7	20629.6	o.
		16409.8	17255.9	18101.0	18944.6	19787.1	20629.9	c.
111		16429.6	17277.6	18133.2	18973.7	19832.7	20682.2	o.
		16425.3	17274.3	18124.4	18973.1	19824.6	20681.8	c.
110		16430.0	17278.1	18133.2	18960.9			o.
		16426.4	17275.5	18124.7	18974.3	19825.3	20682.3	c.
202		16428.7	17274.0	18129.1	18960.9	19801.9	20644.7	o.
		16426.7	17272.8	18119.0	18961.8	19804.7	20647.2	c.
212		16443.2	17293.9	18151.2	18992.5	19842.6	20700.9	o.
		16441.1	17290.1	18139.3	18989.3	19841.5	20700.1	c.
211		16447.8	17297.5	18155.3	18997.3	19852.1		o.
		16444.5	17293.8	18143.2	18993.0	19844.6	20701.6	c.
221		16492.1	17357.7	18227.2	19097.0			o.
		16490.2	17347.7	18208.2	19073.1	19944.6	20822.4	c.
220		16492.3	17357.4	18227.5	19096.9			o.
		16490.1	17347.8	18208.2	19073.1	19044.5	20822.3	c.
303		16452.2	17299.2	18153.9	18986.7	19827.8	20670.7	o.
		16451.6	17297.9	18143.3	18987.2	19829.9	20672.9	c.
313		16465.6	17318.5	18172.7	19014.2	19863.3	20695.0	o.
		16464.7	17313.7	18163.2	19013.6	19866.8	20727.4	c.
312		16475.0	17325.4	18186.3	19880.3			o.
		16471.6	17321.2	18170.9	19021.1	19873.1	20737.7	c.
322		16517.1	17383.0	18254.8	19124.4	20017.6		o.
		16515.8	17373.6	18234.5	19100.3	19973.4	20854.0	c.
321		16517.1	17384.0	18254.6	19125.7			o.
		16515.8	17373.8	18234.7	19100.4	19973.3	20853.5	c.
331		16596.8	17480.3	18368.2	19261.5	20287.3		o.
		16594.2	17462.9	18339.5	19224.7	20120.5	21024.9	c.
330		16596.4	17480.1	18368.1	19261.4			o.
		16592.8	17462.7	18339.4	19224.7	20120.5	21024.8	c.

o.: Observed, from Dubois [2, 5].
c.: Calculated.

With the wavefunctions such obtained, fluorescence lifetimes of the excited state levels were computed for SiH_2 and SiD_2. Very good agreement between the observed and calculated lifetimes was found for most of the vibrational states. The authors were able to predict that, in the absence of dissociation processes, the effect of the Renner–Teller coupling would be to increase the radiative lifetime of levels of SiH_2, with $v_2' \geq 4$, and of those of SiD_2, with $v_2' \geq 8$. This type of lifetime lengthening has been observed both in NH_2 [25] and CH_2 [26].

Finally, spin-orbit coupling to the $^3B_1$ state was included in the calculations. Contamination of the $^1B_1$ state by the triplet state can result in two opposing effects: lifetime lengthening or lifetime shortening. If the triplet levels involved in the coupling are very weakly coupled to the dissociation channel, the effect of this coupling is to produce anomalously long radiative lifetimes, as has been observed in CH_2. On the other hand, if the triplet levels are strongly coupled to the triplet dissociation channel, the lifetime may be greatly shortened. On the basis of the lifetimes of the 0_{00} excited state levels, which are not coupled to the ground state by Renner–Teller or rotational asymmetry effects, it would appear that the radiative lifetime of the $^1B_1$ state is about 1.2 μs as deduced by Fukushima et al. [14]. The shorter-lived levels of SiH_2 are then assumed to be due to coupling to dissociative levels of the triplet state. Since it has been noted that the lifetime shortens when $K_a > 0$, it seems likely that the route to the triplet state is via the Renner–Teller coupling mechanism discussed in this paper.

The onset of the second dissociative channel is due probably to direct coupling of the excited state to the spin-allowed singlet dissociative route to Si (^1D) and $H_2(^1\Sigma_g^+)$. It is this which results in the cutoff at $v_2' = 6$ in the absorption spectrum recorded by Dubois [2, 5]. In the more sensitive laser-induced fluorescence experiments, short-lived levels with $v_2' > 6$ have been observed [8, 14].

5.2 THE SPECTRUM OF PH_2

5.2.1 Effects of Orbital Angular Momentum in PH_2 and PD_2: Renner–Teller and Spin-Orbit Coupling

After A. Alijah and G. Duxbury, *J. Opt. Soc. Am. B*, 11:208, 1994 [27], reproduced by permission from The Optical Society.

Introduction

The absorption spectrum of the $\tilde{A}\,^2A_1$-$\tilde{X}\,^2B_1$ electronic transition of PH_2 was first observed by Ramsay [28] in the flash photolysis of phosphine (PH_3). The spectrum was recognized as an unusual one, because the bending vibrational interval exhibited a minimum value before increasing again. Dixon [29] showed that this behavior is due to the occurrence of a barrier to linearity in the excited state, such that the minimum frequency occurs for the bending state lying closest to the barrier. Following this investigation, Guenebaut et al. [30] measured the low-resolution emission spectrum, which permitted the bending frequency to be measured in both electronic states. The rotational analysis of the 0–0 band by Dixon et al. [31] showed that the electronic transition occurs between two states with markedly different equilibrium angles; in the ground state the angle is close to $92°$, whereas in the excited state, it is $\sim 123°$. The 0–0 band appeared to be unperturbed at the resolution used, and the spin-rotation splittings were analyzed on the basis of coupling between two electronic states derived from a common $^2\Pi_u$ state of linear PH_2. A similar model was used by Pascat et al. [32] to analyze the emission spectra of the 000–020 and the 000–030 bands.

A subsequent analysis by Berthou et al. [20] of the bands involving higher bending levels of the $^2A_1$ state has shown that neither the rotational energy-level pattern nor the spin-rotational interaction is well described by the conventional semirigid-rotor model used for the 0–0 band. The first treatment of this problem by Barrow et al. [33] involved the development of a method for treating the coupling between electronic orbital and nuclear vibrational angular momenta and gave a good explanation of the overall patterns of the vibronic origins and the major spin-rotation constants as a function of vibrational and a-axis rotational angular momenta. This model was subsequently used by Vervloet and Berthou [34] to model the behavior of the excited states of PD_2. However, the use of a so-called damped treatment of the coupling between the $\tilde{A}\,^2A_1$ and $\tilde{X}\,^2B_1$ states precluded the calculation of the effects of close resonances in either PH_2 or PD_2. Since the first analysis by Barrow et al. [33], a wealth of new experimental data had become available [35–40]. This motivated Alijah and Duxbury [27] to apply their new model, which is an extension of Barrow et al. [33], to PH_2, following studies on CH_2 [6, 26], SiH_2 [18], and SiH_2^+ [41]. The extensions include the calculation of rovibronic lifetimes and of perturbations associated with spin-orbit coupling.

Results

The parameters in the model Hamiltonian were adjusted simultaneously for the two electronic states starting from the parameters obtained by Barrow et al. [33]. Simultaneous optimizations permits both the weakly and strongly resonant rovibronic levels to be fitted.

One of the characteristics of Renner–Teller mixing is the observation of anomalously large rotational constants, $\bar{B}$, in the $\tilde{A}^2A_1$ state. This occurs because the strongly resonant pairs of vibronic states are each approximately a 50:50 mixture of the upper and the lower electronic states. In Table 5.13, the observed and calculated vibronic levels and the rotational and spin-orbit coupling parameters of the $\tilde{A}^2A_1$ state are given. It can be seen that, although there are some systematic deviations between the observed and the calculated values, the levels with $v_2' = 6$ and $v_2' = 7$, which exhibit the effects of strong resonances, are much better fitted than in the original model.

Table 5.13: Observed and Calculated Vibronic Levels and Rotational and Spin-Orbit Constants of the $\tilde{A}^2A_1$ State of PH$_2$

v_2'	K	T_{obs}	T_{calc}	Obs-Calc	$\bar{B}_{eff}$		A_v^{SO}	
					(obs)	(calc)	(obs)	(calc)
0	0	18276.6	18276.6	0.0	4.95	4.94	–	–
	1	18297.0	18297.1	−0.1	5.28	5.27	1.2	1.25
					4.62	4.62		
	2	18357.7	18358.2	−0.5	–	4.927	2.4	2.46
	3	18457.6	18458.8	−1.2	–	4.938	3.5	3.59
	4	18594.9	18597.0	−2.1	–	4.953	4.4	4.61
	5	18767.6	18771.0	−3.4	–	4.971	5.4	5.52
	6	18973.0	18978.4	−5.4	–	4.991	6.1	6.29
	7	19209.7	19217.1	−7.4	–	5.015	6.7	6.98
1	0	19225.7	19220.3	5.4	4.96	4.97	–	–
	1	19249.2	18243.6	5.6	5.32	5.82	1.8	1.75
					4.61	4.62		
	2	19318.8	19312.8	6.0	–	4.959	3.3	3.35
	3	19432.4	19425.7	7.1	–	4.974	4.7	4.78
	4	19586.6	19579.4	7.4	–	4.992	6.0	5.97
	5	19778.3	19770.9	6.9	–	5.014	7.0	6.97
	6	20003.9	19997.0	6.9	–	5.037	7.7	7.80
	7	29261.2	20255.0	6.2	–	5.065	8.3	3.48
2	0	20165.4	20159.4	6.1	4.94	4.99	–	–
	1	20193.2	20186.7	6.5	5.30	5.38	2.9	2.64
					4.57	4.62		
	2	20275.1	20266.9	8.2	5.02	5.01	5.2	4.91
	3	20406.2	20396.0	10.2	5.02	5.03	6.9	6.71
	4	20580.6	20569.0	11.5	5.04	5.06	8.3	8.03

Table 5.13: Observed and Calculated Vibronic Levels and Rotational and Spin-Orbit Constants of the $\tilde{A}\,^2A_1$ State of PH_2

v_2'	K	T_{obs}	T_{calc}	Obs-Calc	$\bar{B}_{eff}$ (obs)	$\bar{B}_{eff}$ (calc)	A_v^{SO} (obs)	A_v^{SO} (calc)
	5	20794.4	20781.4	13.0	5.04	5.08	9.0	9.03
	6	21042.7	21029.0	13.6	–	5.11	9.7	9.77
	7	21322.0	21308.4	13.5	–	5.13	10.8	10.37
3	0	21093.8	21091.0	2.8	4.97	5.02	–	–
	1	21128.3	21124.4	3.9	5.37	5.43	4.3	4.42
					4.54	4.62		
	2	21228.0	21220.8	7.2	5.03	5.04	8.3	7.73
	3	21381.6	21371.5	10.1	5.03	5.07	9.8	9.84
	4	21581.1	21569.6	12.5	5.06	5.10	11.2	11.07
	5	21819.6	21805.5	14.1	5.06	5.13	11.7	11.82
	6	22092.1	22077.3	14.8	–	5.16	12.2	12.31
	7	22394.7	22380.4	14.3	–	5.171	12.5	12.80
4	0	22009.7	22011.7	−2.0	4.97	5.03	–	–
	1	22054.5	22055.3	−0.8	5.45	5.43	–	7.79
					–	4.70		
	2	22180.5	22175.6	4.9	5.07	5.08	12.4	12.43
	3	22363.5	22355.1	8.4	5.07	5.11	14.4	14.80
	4	22591.0	22581.1	9.9	5.08	5.14	15.1	15.15
	5	22856.4	22846.0	10.4	5.08	5.18	15.2	15.32
	6	23170.7	23144.4	26.4	–	5.20	14.8	15.31
5	0	22911.0	22917.9	−7.9	4.97	5.04	–	–
	1	22974	22977.6	−3.6	5.48	5.54^c	–	12.68
						4.94		
	2	23136	23133.2	2.8	5.2	5.17^c	20	18.49
	3	23352.9	23349.0	3.9	5.2	5.16	19.7	22.04
	4	23612.9	23608.4	4.5	5.1	5.19	19.0	19.33
	5	23905.7	23904.0	1.6	5.1	5.21	18.8	18.61
6	0	23800.2	23809.7	−9.5	5.03	5.04	–	–
	1	23893	23887.0	6.0	–	5.60^c	–	15.18
						5.84		
	2	24096	24092.1	3.9	5.9	5.38^c	25	25.41
	3	24356	24351.3	1.7	5.6	5.190	25	23.94
	4	24645	24647.0	1.9	–	5.233	26	20.25
7	0	24687.8	24699.1	11.3	5.0	5.07	–	–
	1	–	24756.7	–	–			17.94
	2	25063	25042.6	20.4	6.1	5.80^c	29	17.51
	3	25361	25350.3	10.7	6.4	5.248	26	14.14
						(7.14)		
	4	25684	25681.3	2.7	–	5.294	23	10.24

Table 5.13: Observed and Calculated Vibronic Levels and Rotational and Spin-Orbit Constants of the $\tilde{A}\,^2A_1$ State of PH$_2$

v_2'	K	T_{obs}	T_{calc}	Obs-Calc	$\bar{B}_{eff}$		A_v^{SO}	
					(obs)	(calc)	(obs)	(calc)
8	0	25592.5	25612.0	−19.6	5.0	5.12		−

In order to show the success of the semirigid-bender treatment of dihydrides, such as PH$_2$, we show the vibronic levels of PD$_2$, calculated with the parameters derived from the fit to PH$_2$, in Table 5.14. It can be seen that the fit is quite good and accounts for the observed pattern of most of the vibronic and K-type structure and for the vibrational and K-rotational dependence of the rotation constants. The shift of −20 cm^{-1} between the observed and the calculated levels is probably associated with the isotopic dependence of the difference between the zero-point energies of the stretching vibrations in the ground and the excited states, which is not accounted for in the model.

Table 5.14: Observed and Calculated Vibronic Levels and Rotational and Spin-Orbit Constants of the $\tilde{A}\,^2A_1$ State of PD$_2$

v_2'	K	T_{obs}	T_{calc}	Obs-Calc	$\bar{B}_{eff}$		A_v^{SO}	
					(obs)	(calc)	(obs)	(calc)
1	0	18960.0	18981.7	−21.7		2.49	−	−
	4	19144.8	19164.7	−19.9	2.55	2.51	3.21	2.97
	6	19367.0	19385.4	−12.7	2.50	2.52	4.02	4.13
	7	19512.1	19524.9	−12.8	2.41	2.53	4.57	4.63
	8	19666.9	19682.3	−15.4	2.53	2.54	4.96	5.06
2	0	19645.5	19662.2	−16.8	2.49	2.51	−	−
	1	19659	19673.1	−16.1	2.35	2.33	$(1.00)^b$	1.02
					2.66	2.68		
	2	19698.3	19713.2	−14.9	2.46	2.51	(2.0)	1.99
	3	19761.6	19776.1	−14.5	2.51	2.51	3.01	2.88
	4	19849.3	19862.6	−13.3	2.52	2.52	3.75	3.68
	5	19960.5	19971.4	−10.9	2.50	2.53	4.65	4.36
	6	20091.7	20101.1	−9.4	2.52	2.54	5.07	4.97
	7	20242.5	20250.6	−8.1	2.48	2.54	5.60	5.49
	8	20412.4	20418.3	−5.9	2.49	2.55	5.87	5.93
3	0	20326.0	20339.7	−13.7	2.46	2.52	−	−
	1	20340.4	20354.1	−14.9	2.25	2.33	1.38	1.40
					2.62	2.70		
	2	20384.0	20396.8	−12.8	2.51	2.52	2.93	2.69
	3	20455.7	20466.6	−10.9	2.52	2.53	3.88	3.81
	4	20553.3	20561.7	−8.4	2.52	2.53	4.87	4.73

Table 5.14: Observed and Calculated Vibronic Levels and Rotational and Spin-Orbit Constants of the $\tilde{A}\,^2A_1$ State of PD_2

v_2'	K	T_{obs}	T_{calc}	Obs-Calc	$\bar{B}_{eff}$ (obs)	(calc)	A_v^{SO} (obs)	(calc)
	5	20674.0	20680.5	−6.5	2.52	2.54	5.76	5.49
	6	20817.2	20821.1	−3.9	2.52	2.55	6.23	6.10
	7	20980.2	20981.7	−1.5	2.51	2.56	6.69	6.61
	8	21162.0	21160.9	1.1	2.52	2.57	7.11	7.01
4	0	20999.0	21013.0	−14.0	2.48	2.52	−	−
	1	21016.6	21029.5	−12.9	2.28	2.34	2.27	2.14
					2.64	2.71		
	2	21065.9	21098.2	−12.3	2.48	2.53	4.13	3.98
	3	21146.8	21156.8	−10.0	2.53	2.55	5.61	5.38
	4	21255.6	21262.7	−7.1	2.54	2.55	6.54	6.39
	5	21389.0	21393.3	−4.3	2.52	2.56	7.10	7.13
	6	21545.0	21546.2	−1.2	2.53	2.57	7.85	7.67
	7	21721.5	21719.4	2.1	2.51	2.58	8.23	8.07
	8	21916.0	21911.1	4.9	2.50	2.59	8.53	8.40
5	0	21665.5	21680.5	−18.9	2.45	2.53	−	−
	1	21685.8	21706.2	−14.4	2.26	2.34	(3.5)	3.64
						2.73		
	2	21743.3	21757.3	−14.0	2.57	2.55	(5.45)	6.36
	3	21837.4	21847.5	−10.1	2.55	2.56	7.49	8.03
	4	21959.1	21966.7	−7.6	2.57	2.57	8.61	8.98
	5	22107.7	22111.1	−3.4	2.55	2.58	9.37	9.51
	6	22278.4	22278.0	0.4	2.51	2.58	9.61	9.81
	7	22467.0	22464.9	2.1	2.53	2.61	9.90	10.10
6	0	22322.0	22340.4	−18.4	2.32	2.53	−	−
	1	−	22365.2	−	−	2.36	(5.2)	6.63
						2.75		
	2	22417.5	22434.6	−17.1	2.67	2.58	8.69	10.51
	3	22526.5	22540.2	−13.7	2.65	2.60	11.04	12.15
	4	22665.9	22675.4	−9.5	2.62	2.61	11.66	12.68
	5	22829.5	22835.7	−6.2	2.58	2.62	11.95	12.70
	6	23014.0	23017.9	−3.9	2.58	2.63	12.02	12.58
	7	23219.0	23219.7	−0.7	2.55	2.63	11.97	12.37
7	0	22971.5	22990.3	−18.8	2.32	2.53	−	−
	1	−	23023.7	−19.6	−	2.42	−	11.63
						2.76		
	2	23091.7	23111.3	−19.6	2.67	2.65	12.22	16.48
	3	23218.8	23236.5	−17.7	2.65	2.67	15.05	17.45
	4	23376.4	23390.4	−14.0	2.62	2.68	15.01	17.18
	5	23558.0	23568.2	−10.2	2.58	2.67	15.02	16.49

Table 5.14: Observed and Calculated Vibronic Levels and Rotational and Spin-Orbit Constants of the $\tilde{A}\,^2A_1$ State of PD_2

v_2'	K	T_{obs}	T_{calc}	Obs-Calc	$\bar{B}_{eff}$ (obs)	(calc)	A_v^{SO} (obs)	(calc)
	6	23758.9	23767.1	−8.2	2.58	2.68	14.51	15.82
	7	23975.9	23984.6	−8.7	2.55	2.68	14.03	14.97
8	0	23605.5	23628.0	−22.5	2.50	2.53	−	−
	1	−	23675.7	−	−	2.77	−	17.89
	3	23915.6	23937.3	−21.7	2.62	2.77	19.38	22.59
	4	24092.6	24112.1	−19.5	2.66	2.76	18.67	21.48
	5	24292.0	24308.9	−16.9	2.60	2.76	17.59	19.94
	6	24506.7	24525.2	−18.5	2.70	2.76	14.83	18.81
9	0	24233.0	24254.1	−21.1	2.55	2.53	−	
	3	24615.6	24641.0	−25.4	2.62	2.94	24.83	25.05
	4	24814.0	24838.1	−24.1	2.81	2.93	22.76	23.23
10	0	24857.0	24877.0	−20.0	2.6	2.54	−	−
11	0	25493.0	25513.1	−20.1	2.6	2.56	−	−

The effect of the resonances can be seen most vividly by examination of the variation of the spin-orbit coupling with the vibronic state by use of an alternative method of plotting that is from Jungen and Merer [42]. In these plots the calculated spin-orbit interaction in both the $^2A_1$ and the $^2B_1$ states is plotted as a function of vibronic level energy. This is shown in Figure 5.27 for $K_a = 1$. All molecules of this type exhibit a sawtooth or mirror pattern of the induced spin-orbit coupling pattern in vibronic levels of the lower and upper electronic states that lie close to the barrier to linearity. If the resonant terms are ignored, the calculated damping of the angular momentum is far too slow, as is shown in these diagrams. The observation of such sawtooth behavior shows that a large amount of electronic orbital angular momentum is generated. The complications associated with the generation of large amounts of angular momentum in vibronic levels of PH_2 that lie close to the barrier to linearity are probably the reason that many of the rovibronic levels of $v_2' = 6$ and $v_2' = 6$ with $K_a = 1$, and all those of $v_2' = 7$ with $K_a = 1$, have not been assigned so far.

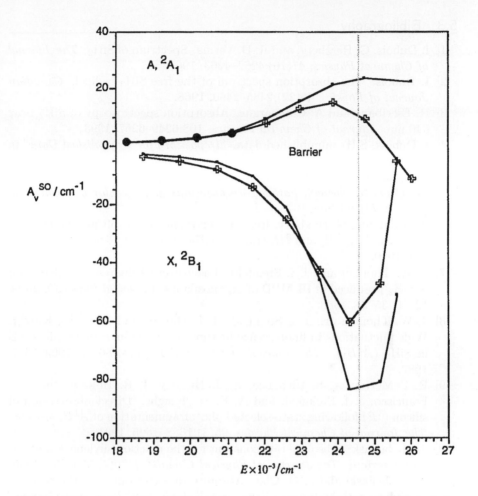

Figure 5.27: Spin-orbit splitting of the vibronic levels with $K = 1$. Filled circles, observed; open pluses, calculated; filled squares, calculated with resonant terms ignored.

5.3 Bibliography

[1] I. Dubois, G. Herzberg, and R. D. Verma. Spectrum of SiH$_2$. *The Journal of Chemical Physics*, 47(10):4262–4263, 1967.

[2] I. Dubois. The absorption spectrum of the free SiH$_2$ radical. *Canadian Journal of Physics*, 46(22):2485–2490, 1968.

[3] R. Escribano and A. Campargue. Absorption spectroscopy of SiH$_2$ near 640 nm. *Journal of Chemical Physics*, 108:6249–6257, 1998.

[4] I. Dubois. SiH$_2$ unpublished data. *"Depository for Unpublished Data" in the National Science Library of Ottawa, Canada*, 1975.

[5] I. Dubois, G. Duxbury, and R. N. Dixon. Renner effect in SiH$_2$. *Journal of the Chemical Society, Faraday Transactions 2: Molecular and Chemical Physics*, 71:799–806, 1975.

[6] A. Alijah and G. Duxbury. Renner–Teller and spin-orbit interactions between the $^1A_1$, $^1B_1$ and $^3B_1$ states of CH$_2$. *Molecular Physics*, 70(4):605–622, 1990.

[7] J. W. Thoman and J. I. Steinfeld. Laser-excited fluorescence detection of SiH$_2$ produced in IR MPD of organosilanes. *Chemical Physics Letters*, 124(1):35–38, 1986.

[8] J. W. Thoman Jr., J. I. Steinfeld, R. I. McKay, and A. E. W. Knight. Wide fluctuations in fluorescence lifetimes of individual rovibronic levels in SiH$_2$ ($\tilde{A}^1B_1$). *The Journal of Chemical Physics*, 86(11):5909–5917, 1987.

[9] R. I. McKay, A. S. Uichanco, A. J. Bradley, J. R. Holdsworth, J. S. Francisco, J. I. Steinfeld, and A. E. W. Knight. Direct observation of silicon (^3P) following state-selected photofragmentation of $\tilde{A}^1B_1$ silylene. *The Journal of Chemical Physics*, 95(3):1688–1695, 1991.

[10] J. M. Jasinski. Absolute rate constant for the reaction silylene + molecular deuterium. *The Journal of Physical Chemistry*, 90(4):555–557, 1986.

[11] J. M. Jasinski and J. O. Chu. Absolute rate constants for the reaction of silylene with hydrogen, silane, and disilane. *The Journal of Chemical Physics*, 88(3):1678–1687, 1988.

[12] J. J. O'Brien and G. H. Atkinson. Role of silylene in the pyrolysis of silane and organosilanes. *The Journal of Physical Chemistry*, 92(20):5782–5787, 1988.

[13] K. Tachibana, T. Shirafuji, and Y. Matsui. Measurement of SiH$_2$ densities in an RF-discharge silane plasma used in the chemical vapor deposition of hydrogenated amorphous silicon film. *Japanese Journal of Applied Physics*, 31(8R):2588–2591, 1992.

[14] M. Fukushima, S. Mayama, and K. Obi. Jet spectroscopy and excited state dynamics of SiH$_2$ and SiD$_2$. *The Journal of Chemical Physics*, 96(1):44–52, 1992.

[15] M. Fukushima and K. Obi. Rotational analysis of the SiD$_2$ $\tilde{A}^1B_1$ – $\tilde{X}^1A_1$ transition observed in a jet. *The Journal of Chemical Physics*, 100(9):6221–6227, 1994.

[16] A. Campargue, D. Romanini, and N. Sadeghi. Measurement of density in a discharge by intracavity laser absorption spectroscopy and cw cavity ring-down spectroscopy. *Journal of Physics D: Applied Physics*, 31(10):1168–1175, 1998.

[17] C. Yamada, H. Kanamori, E. Hirota, N. Nishiwaki, N. Itabashi, K. Kato, and T. Goto. Detection of the silylene ν_2 band by infrared diode laser kinetic spectroscopy. *The Journal of Chemical Physics*, 91(8):4582–4586, 1989.

[18] G. Duxbury, A. Alijah, and R. R. Trieling. Renner–Teller and spin-orbit interactions in SiH_2. *The Journal of Chemical Physics*, 98(2):811–825, 1993.

[19] J. O'Brien and G. Atkinson. Detection of the SiH_2 radical by intracavity laser absorption spectroscopy. *Chemical Physics Letters*, 130(4):321–329, 1986.

[20] J. M. Berthou, B. Pascat, H. Guenebaut, and D. A. Ramsay. Rotational analysis of bands of the transition of PH_2. *Canadian Journal of Physics*, 50(19):2265–2276, 1972.

[21] H. Petek, D. J. Nesbitt, D. C. Darwin, and C. Bradley Moore. Visible absorption and magnetic-rotation spectroscopy of $^1CH_2$: The analysis of the $\tilde{b}^1B_1$ state. *The Journal of Chemical Physics*, 86(3):1172–1188, 1987.

[22] H. Petek, D. J. Nesbitt, C. Bradley Moore, F. W. Birss, and D. A. Ramsay. Visible absorption and magnetic-rotation spectroscopy of $^1CH_2$: Analysis of the $^1A_1$ state and the $^1A_1 - {}^3B_1$ coupling. *The Journal of Chemical Physics*, 86(3):1189–1205, 1987.

[23] J. S. Francisco, R. Barnes, and J. W. Thoman Jr. Dissociation dynamics of low-lying electronic states of SiH_2. *The Journal of Chemical Physics*, 88(4):2334–2341, 1988.

[24] C. M. van Zoeren, J. W. Thoman, J. I. Steinfeld, and M. Rainbird. Production of silicon ($^1D_2$) from electronically excited silylene. *The Journal of Physical Chemistry*, 92(1):9–11, 1988.

[25] Ch. Jungen, K.-E. J. Hallin, and A. J. Merer. Orbital angular momentum in triatomic molecules. II. Vibrational and K-type rotational structure, and intensity factors in the $A^2A_1 - X^2B_1$ transitions of NH_2 and H_2O^+. *Molecular Physics*, 40(1):25–63, 1980.

[26] I. García-Moreno, E. R. Lovejoy, C. Bradley Moore, and G. Duxbury. Radiative lifetimes of CH_2 (b^1B_1). *The Journal of Chemical Physics*, 98(2):873–882, 1993.

[27] A. Alijah and G. Duxbury. Effects of orbital angular momentum in PH_2 and PD_2: Renner–Teller and spin-orbit coupling. *Journal of the Optical Society of America B*, 11(1):208–218, January 1994.

[28] D. A. Ramsay. Effects of orbital angular momentum in PH_2 and PD_2: Renner–Teller and spin-orbit coupling. *Nature*, 178:374–375, 1956.

[29] R. N. Dixon. Higher vibrational levels of a bent triatomic molecule. *Transactions of the Faraday Society*, 60:1363–1368, 1964.

[30] H. Guenebaut, B. Pascat, and J. M. Berthou. Sur les emissions du radical

PH_2. *Journal de Chimie Physique*, 62:867–867, 1965.

[31] R. N. Dixon, G. Duxbury, and D. A. Ramsay. Rotational analysis of the 0-0 band of the $^2A_1 - ^2B_1$ electronic transition of PH_2. *Proceedings of the Royal Society of London A: Mathematical, Physical and Engineering Sciences*, 296(1445):137–160, 1967.

[32] B. Pascat, J. M. Berthou, J. C. Prudhomme, H. Guenebaut, and D. A. Ramsay. Rotational analysis of 000-020 and 000-030 bands in transition $^2A_1 - ^2B_1$ of triatomic radical PH_2. *Journal de Chimie Physique et de Physico-Chimie Biologique*, 65(11-1):2022–2029, 1968.

[33] T. Barrow, R. N. Dixon, and G. Duxbury. The Renner effect in a bent triatomic molecule executing a large amplitude bending vibration. *Molecular Physics*, 27(5):1217–1234, 1974.

[34] M. Vervloet and J. M. Berthou. Etude du système A-X de PD_2. *Canadian Journal of Physics*, 54(13):1375–1382, 1976.

[35] R. E. Huie, N. J. T. Long, and B. A. Thrush. Laser induced fluorescence of the PH_2 radical. *Journal of the Chemical Society, Faraday Transactions 2: Molecular and Chemical Physics*, 74:1253–1262, 1978.

[36] C. Nguyen Xuan and A. Margani. Dynamics and spectroscopy of $PH_2(\tilde{A}^2A_1)$. *The Journal of Chemical Physics*, 93(1):136–146, 1990.

[37] M. Kakimoto and E. Hirota. Hyperfine structure of the PH_2 radical in $\tilde{X}^2B_1$ and $\tilde{A}^2A_1$ from intermodulated fluorescence spectroscopy. *Journal of Molecular Spectroscopy*, 94(1):173–191, 1982.

[38] A. R. W. McKellar. Mid-infrared laser magnetic resonance spectroscopy. *Faraday Discussions of the Chemical Society*, 71:63–76, 1981.

[39] D. Baugh, B. Koplitz, Z. Xu, and C. Wittig. PH_2 internal energy distribution produced by the 193 nm photodissociation of PH_3. *The Journal of Chemical Physics*, 88(2):879–887, 1988.

[40] I. R. Lambert, G. P. Morley, D. H. Mordaunt, M. N. R. Ashfold, and R. N. Dixon. Near ultraviolet photolysis of phosphine studied by H atom photofragment translational spectroscopy. *Canadian Journal of Chemistry*, 72(3):977–984, 1994.

[41] D. I. Hall, A. P. Levick, P. J. Sarre, C. J. Whitham, A. Alijah, and G. Duxbury. Faraday communications. High-resolution electronic spectroscopy and predissociation dynamics of SiH_2^+ in high-K states. *Journal of the Chemical Society, Faraday Transactions*, 89:177–178, 1993.

[42] Ch. Jungen and A. J. Merer. Orbital angular momentum in triatomic molecules. I. A general method for calculating the vibronic energy levels of states that become degenerate in the linear molecule (the Renner–Teller effect). *Molecular Physics*, 40(1):1–23, 1980.

CHAPTER 6

Third Row Dihydrides

CONTENTS

6.1   THE SPECTRUM OF GeH_2

6.1.1   Introduction

The first electronic spectra of germylene were reported by Saito and Obi in 1993, using laser-induced fluorescence (LIF) spectra of jet-cooled GeH_2 and GeD_2, produced by photolysis of phenylgermane [1, 2]. It was later found in subsequent work in Kentucky that strong LIF signals could be obtained using the discharge jet technique with germane as the precursor [3].

     More recently, Campargue and Escribano [4] reported the room-temperature absorption spectrum of the central part of the 2^1_0, i.e.,

$v_2'' = 0 - v_2' = 1$, band of GeH_2 recorded using intracavity laser absorption spectroscopy (ICLAS). Although their spectra showed well-resolved germanium isotope splittings, only transitions to $K_a' = 0$ and 1 levels were observed. From this they obtained excited state rotational constants for the five GeH_2 isotopomers and estimated the r_e' structure. GeH_2 undergoes a rotationally induced predissociation in the excited state, much like that of the SiH_2 radical; see Fukushima et al. [5].

6.1.2   The Excited State Dynamics of the $\tilde{A}\,^1B_1$ States of GeH_2 and GeD_2 Radicals

K. Saito and K. Obi, Department of Chemistry, Tokyo Institute of Technology, Japan [6].

Laser-induced fluorescence excitation spectra of the $\tilde{A}\,^1B_1 - \tilde{X}\,^1A_1$ transition of germylene radicals have been measured in a supersonic free jet. The spectra consist of progressions in the bending mode, that is, $(0v_2'0) - (000)$, $v_2' = 0 - 4$ for GeH_2 and $v_2' = 0 - 6$ for GeD_2. The spectrum of GeH_2 is shown in Figure 6.1. Rotational assignments have been made. The only lines appearing belong to the $K_a' = 0 - K_a'' = 1$ subbands, except for the $(000) - (000)$ band of GeD_2, where the $^rR_0(0)$ line is also seen; for the higher v_2' levels of both isotopologues, the bands consist of a single line, $^pP_1(1)$. This suggests the presence of heterogeneous predissociation to $Ge(^3P) + H_2$ in the $\tilde{A}^1B_1$ state. The fluorescence lifetimes have been measured for single rovibronic levels of the $\tilde{A}^1B_1$ state of both radicals. The fluorescence lifetimes of the rotational levels of the $\tilde{A}^1B_1$ state decrease with increasing vibrational energy. A second predissociation channel to $Ge(^1D) + H_2$ opens between $v_2' = 4$ of GeH_2 and $v_2' = 7$ of GeD_2. The heat of formation of GeH_2 is estimated to be between 19053 and 19178 cm^{-1}.

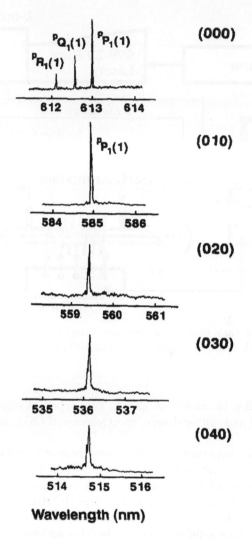

Figure 6.1: LIF excitation spectra for the $\tilde{A}\,^1B_1\,(0v_2'0) - \tilde{X}\,^1A_1\,(000)$ bands, with $v_2' = 0 - 4$, of GeH_2. Reprinted from Saito and Obi, *Chem. Phys.*, 187:381, 1994 [6]. Copyright (1994), with permission from Elsevier.

6.1.3 Laser Optogalvanic and Jet Spectroscopy of GeH_2

T. C. Smith and D. Clouthier, Department of Chemistry, University of Kentucky, Lexington, Kentucky, and W. Sha and A. G. Adam, Department of Chemistry, University of New Brunswick, Fredericton, New Brunswick, Canada [7].

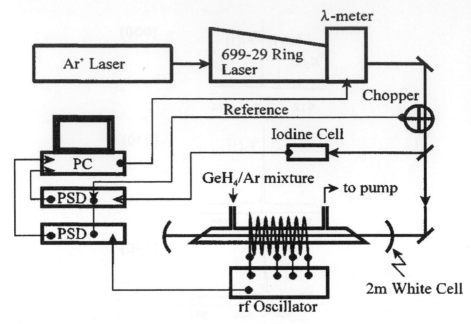

Figure 6.2: Schematic diagram of the laser optogalvanic (LOG) apparatus. Reproduced from Smith et al., *J. Chem. Phys.*, 113:9567, 2000 [7], with the permission of AIP Publishing.

In the studies by Smith et al. [7], two types of experimental methods were used: 1. Laser optogalvanic spectroscopy (LOG), and 2. Fluorescence spectroscopy.

The room-temperature absorption spectrum of germylene was studied by LOG spectroscopy. Germylene was produced in radio frequency discharge through a mixture of GeH_2 and argon. A schematic representation of the LOG experiment is given in Figure 6.2. The 10 MHz radio frequency used in this experiment was based on that of Lyons et al. [8], Vasudev and Zarek [9], and modified using a solid-state circuit by May and May [10]. Low and hi-pass filters on the lock-in amplifier were used to reduce the noise. A time constant of 3 s was used to average the signal, and an output filter of 1 s was used before the signal was sent to the computer that controlled the data collection and scanning of the laser. A cw ring dye laser was used to record the LOG spectra. Two laser dyes were used in studies of the 0_0^0 (610 nm) and 2_0^1 (589 nm) bands. The LOG spectra were calibrated by simultaneously recording the LIF spectrum of iodine vapour. The accuracy was about $0.003\,\mathrm{cm}^{-1}$ or 90 MHz. Typical linewidths in the LOG spectrum were about $0.03\,\mathrm{cm}^{-1}$ FWHM.

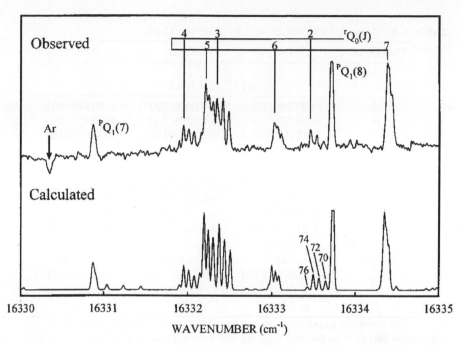

Figure 6.3: A portion of the room-temperature LOG absorption spectrum of GeH_2. The top trace shows the central portion of the 0_0^0 band with leaders marking some of the rotational assignments. The bottom trace is a simulated spectrum using the rotational constants determined from the analysis of the LOG spectrum. Reproduced from Smith et al., *J. Chem. Phys.*, 113:9567, 2000 [7], with the permission of AIP Publishing.

As shown in Figure 6.3, a variety of weak, complex features were found near the band centre, along with a few weak argon lines, which exhibited the opposite phase. Some of the features in the spectrum showed the typical five-line germanium isotope pattern, whereas the others have no discernable splittings. Most of the strong features could be assigned as the $^rQ_0$ and $^PQ_1$ branches, terminating on the $K_a = 0$ and 1 levels in the excited state. The low J lines showed resolved isotope splittings, whereas the high J transitions are usually blended together. Smith and colleagues could not identify any transitions in the LOG spectra with $K_a' > 1$. It appeared that the higher K_a excited state GeH_2 rotational levels are so short lived, due to the rotationally induced predissociation to $Ge+H_2$, that they are not detectable by absorption spectroscopy.

The 0_0^0 bands of the most abundant isotopologues, $^{70}GeH_2$, $^{72}GeH_2$, and $^{74}GeH_2$, were analysed and rotational parameters extracted. These parameters are collected in Table 6.1.

Table 6.1: Molecular Constants (In cm^{-1}) of the GeH_2

Parameter	$^{70}GeH_2$	$^{72}GeH_2$	$^{74}GeH_2$
	$\tilde{X}^1 A_1\, v'' = 0$ level		
A	$7.02730(41)^a$	$7.02006(50)$	$7.01876(65)$
B	$6.53075(29)$	$6.53228(33)$	$6.52915(27)$
C	$3.33352(35)$	$3.33708(19)$	$3.33030(15)$
$10^3\Delta_K$	1.35	1.35	$1.350(52)$
	$\tilde{A}^1 B_1\, v' = 0$ level		
A	$16.3843(11)$	$16.3687(13)$	$16.3629(11)$
B	$4.52159(36)$	$4.52428(35)$	$4.52227(14)$
C	$3.46025(36)$	$3.46458(24)$	$3.45826(22)$
$10^3\Delta_K$	-2.87	-2.87	$-2.868(49)$
$10^5\Delta_J$	3.44	3.44	$3.444(48)$
T_0	$16\,325.6786(13)$	$16\,325.6081(14)$	$16\,325.5393(10)$
σ^b	0.0034	0.0049	0.0048
$\#^c$	15	19	30

[a] The numbers in parentheses are three standard deviations and are right justified to the last digit on the line; sufficient additional digits are quoted to reproduce the original data to full accuracy. Values without a standard deviation were constrained in the fit.
[b] The overall standard deviation of fit.
[c] The number of transitions fitted.

LIF and wavelength resolved fluorescence spectra of GeH_2 and GeD_2 were recorded under jet-cooled conditions using a pulsed discharge jet apparatus [11]. Although germane is a good precursor for germylene, in a pulsed discharge jet, it was discovered that monochlorogermane is easier to handle and gives stronger LIF spectra.

Low-resolution LIF survey spectra of GeD_2 were recorded in the 620–505 nm region. The D_3GeCl precursor produced strong GeD_2 in this region, without interferences from other fluorescent species. Twelve bands were found, all originating from the lowest vibrational level in the ground state. Seven were assigned as the 2_0^n, $n = 0-6$ bending progression of Saito and Obi [6]. Several high-resolution spectra of the $^pP_1(1)$ transition of each band was recorded at high resolution. Each $^pP_1(1)$ rovibronic transition shows five features due to the germanium isotopes of significant natural abundance, ($^{76}Ge = 7.8\%$, $^{74}Ge = 36.5\%$, $^{73}Ge = 7.8\%$, $^{72}Ge = 27.4\%$, $^{70}Ge = 20.5\%$). The measured $^pP_1(1)$ transition frequencies for the various vibronic bands are listed in Table 6.2.

Table 6.2: Measured $^PP_1(1)$ Isotopic Line Frequencies (In cm^{-1}) For Various Vibronic Bands of GeD$_2$

Band	^{76}GeD$_2$	^{74}GeD$_2$	^{73}GeD$_2$	^{72}GeD$_2$	^{70}GeD$_2$
2_0^1	16878.234	16878.612	16878.809	16879.012	16879.441
2_0^2	17437.595	17438.279	17438.621	17439.979	17439.749
2_0^3	17995.187	17996.161	17996.681	17997.187	17998.285
2_0^4	18550.761	18552.034	18552.704	18553.375	18554.784
2_0^5	19104,032	19105.589	19106.402	19107.227	19108.934
2_0^6	19654.740	19656.559	19657.507	19658.481	19660.507
1_0^1	17621.542	17621.775	17621.903	17622.023	17622.287
$1_0^1 2_0^1$	18174.153	18174.688	18174.968	18175.262	18175.868
$1_0^1 2_0^2$	18725.534	18726.831	18726.831	18727.281	18728.228
$1_0^1 2_0^3$	19275.562	19276.712	19277.317	19277.931	19279.209
$1_0^1 2_0^4$	19824.050	19825.491	19826.250	19827.017	19828.638

The vibronic band origins for the ^{74}GeD$_2$ isotopomer were estimated by adding the ground state $1_{1,0} - 0_{0,0}$ interval (6.885 cm^{-1}) determined from the rotational analysis of the ^{74}GeD$_2$ 0_0^0 band spectrum to the measured $^PP_1(1)$ rovibronic transition frequencies; the results are summarized in Table 6.3. The vibronic band origins were fitted to the standard anharmonic expression to obtain the excited state vibrational frequencies.

$$\nu = T_{00} + \omega_1^0 v_1 + \omega_2^0 v_2 + x_{22}^0 v_2^2 + x_{12}^0 v_1 v_2 \tag{6.1}$$

T_{00} was fixed at the value (16324.3387 cm^{-1}) obtained from the rotational analysis of the 0_0^0 band of ^{74}GeD$_2$. The optimised parameter values, with 1σ standard error in parentheses, are $\omega_1^0 = 1303.73(86)$, $\omega_2^0 = 562.41(54)$, $x_{22}^0 = -0.969(76)$, and $x_{12}^0 = -7.59(31)$.

Table 6.3: Vibrational Assignments and Band Origins (In cm^{-1}) of the Observed Vibronic Bands of $^{74}GeD_2$

Assignment	Band origin[a]	Obs-Calc[b]
0^0_0	16 324.339	0
2^1_0	16 885.497	−0.29
2^2_0	17 445.164	−0.13
2^3_0	18 003.046	0.18
2^4_0	18 558.918	0.42
2^5_0	19 112.473	0.28
2^6_0	19 663.443	−0.51
1^1_0	17 628.660	0.60
$1^1_0 2^1_0$	18 181.572	−0.35
$1^1_0 2^2_0$	18 733.272	−0.56
$1^1_0 2^3_0$	19 283.597	−0.21
$1^1_0 2^4_0$	19 832.375	0.53

[a]Band origin of the 0^0_0 band from Table 6.4; other bands from $^pP_1(1)$ line $+6.885 cm^{-1}$ ($1_{1,0} - 0_{0,0}$ interval in ground state).
[b]Calculated using Equation (6.1).

For the isotopologues $^{70}GeD_2$, $^{72}GeD_2$, $^{74}GeD_2$ and $^{76}GeD_2$, the the high-resolution 0^0_0 bands were analysed and rotational parameters determined. They are presented in Table 6.4.

Wavelength-resolved fluorescence spectra of GeH_2 and GeD_2 were collected in the 550–850 nm region. An example spectrum, showing the 0^0_0 emission of jet-cooled GeD_2, is presented in Figure 6.4. Analysis of these data led to estimations of band origins for $^{74}GeH_2$ and $^{74}GeD_2$.

Table 6.4: Molecular Constants (In cm^{-1}) of the GeD$_2$[a]

Parameter	^{70}GeD$_2$	^{72}GeD$_2$	^{74}GeD$_2$	^{76}GeD$_2$
	$\tilde{X}^1 A_1\, v'' = 0$ level			
A	3.63141(26)	3.62156(37)	3.61767(44)	3.60930(49)
B	3.27323(12)	3.27307(11)	3.27313(12)	3.27302(13)
C	1.700415(62)	1.699379(55)	1.698058(61)	1.697205(70)
$10^3 \Delta_K$	1.173(20)	0.662(39)	0.870(45)	0.538(51)
	$\tilde{A}^1 B_1\, v' = 0$ level			
A	8.37179(35)	8.36154(28)	8.34752(32)	8.33381(34)
B	2.269 632(69)	2.269 559(49)	2.269 216(56)	2.270 223(69)
C	1.757 279(79)	1.757 203(64)	1.756 460(69)	1.755 923(86)
$10^3 \Delta_K$	6.600(37)	6.815(30)	6.622(35)	6.518(37)
T_0	16 324.5086(78)	16 324.4123(71)	16 324.3387(84)	16 324.2614(88)
σ^b	0.0074	0.0059	0.0072	0.0063
#^{c}	72	69	75	56

[a]The numbers in parentheses are three standard deviations and are right justified to the last digit on the line; sufficient additional digits are quoted to reproduce the original data to full accuracy.
[b]The overall standard deviation of fit.
[c]The number of transitions fitted.

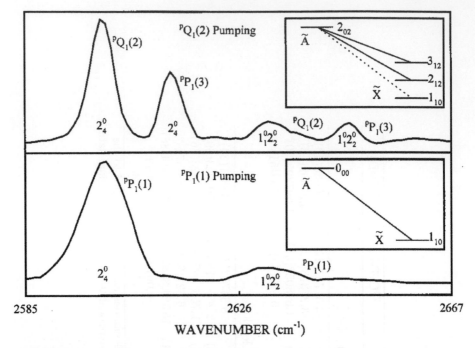

Figure 6.4: A portion of the 0_0^0 band emission of jet-cooled GeD$_2$. The top trace was recorded by exciting the $^PQ_1(2)$ transition, while the bottom trace was recorded by exciting the weaker $^PP_1(1)$ transition. The insets show the allowed rovibronic transitions in each case. Reproduced from Smith et al., *J. Chem. Phys.*, 113:9567, 2000 [7], with the permission of AIP Publishing.

The rotational constants obtained in this work have permitted a new derivation of the structural parameters at equilibrium for the two electronic states; see Table 6.5. Fundamental frequencies are collected in Table 6.6.

Table 6.5: Comparison of Experimental and Ab Initio Geometric Parameters

| | $\tilde{X}^1 A_1$ | | $\tilde{A}^1 B_1$ | |
Structure	r(Ge-H)/Å	θ(H-Ge-H)/°	r(Ge-H)/Å	θ(H-Ge-H)/°
r_0^a	1.5934(2)	91.28(1)	1.5564(5)	123.02(2)
$r_0^{\epsilon a}$	1.5883(9)	91.22(1)	1.5471(6)	123.44(2)
r_0^b	1.591(7)	91.2(8)	1.553(12)	123.4(19)
r_e^c			1.5422(13)	122.82(20)
r_e^d	1.591	91.4	1.553	122.1

[a]Experimental value from this work [7]. Values in parentheses are three standard deviations.
[ϵa] Most accurate experimental value from this work, using Rudolphs's method.
[b]Experimental value, Reference [3].
[c]Experiment plus ab initio theory, estimated structure, Reference [4].
[d]Ab initio plus semirigid bender theory, Reference [12].

Table 6.6: Experimental and Theoretical Vibration Frequencies (in cm^{-1}) For The $\tilde{X}^1 A_1$ and $\tilde{A}^1 B_1$ States of GeH$_2$ and GeD$_2$

| Electronic | Experimental[a] | | | | Theory | | | |
State	ν_1	ν_2	ν_3	Ref.	ν_1	ν_2	ν_3	Ref.
			^{74}GeH$_2$					
$\tilde{X}^1 A_1$	1887	920	1864	[13]	1857	923	1866	[12][c]
	1856	916	...	this work [7]	1840	913	1840	[3][b]
$\tilde{A}^1 B_1$	...	~780	...	[1,6]	1864	860	2011	[14]
	1798	783	...	[3][b]	1809	783	1909	[3][b]
			^{74}GeD$_2$					
$\tilde{X}^1 A_1$	1327	657	1338	[13]	1339	657	1352	[12]
	1335	657	...	this work [7]				
$\tilde{A}^1 B_1$	...	~563	...	[1,6]				
	1304	561	...	this work [7]				

[a]Fundamental vibrational frequencies.
[b]Values obtained from previous LIF study.
[c]Harmonic frequencies from a fitted potential surface.

6.2 THE SPECTRUM OF AsH_2

6.2.1 Introduction

The electronic spectrum of AsH_2 was studied using photographic flash photolysis techniques by Dixon, Duxbury, and Lamberton in 1967 [15]. The authors found a simple progression of bands in the 500–400 nm region, which they assigned as the $2_0^n (n = 0 - 6)$, i.e., $v_2'' = 0 - v_2' = n$, vibronic transitions. Rotational analysis of the 2_0^1, 2_0^2, and 2_0^3 bands showed that the electronic transition is $^2A_1 - {}^2B_1$ with rather large spin splittings in the excited state [16]. The ground state has been extensively studied in recent years. Saito and his colleagues [17, 18] obtained the microwave spectra of AsH_2 and AsD_2, and Hughes, Brown, and Evenson recorded the rotational spectra of the AsH_2 radical by Far-Infrared laser magnetic resonance [19].

6.2.2 The Analysis of a $^2A_1 - {}^2B_1$ Electronic Band System of the AsH_2 and AsD_2 Radicals

R. N. Dixon, G. Duxbury, and H. M. Lamberton, *Proc. Roy. Soc. A.*, 305:271, 1968.

Key rotational transitions and vibrational intervals (in cm^{-1}) identified in the absorption spectrum of AsH_2 by Dixon, Duxbury, and Lamberton [20] are shown in Figure 6.5.

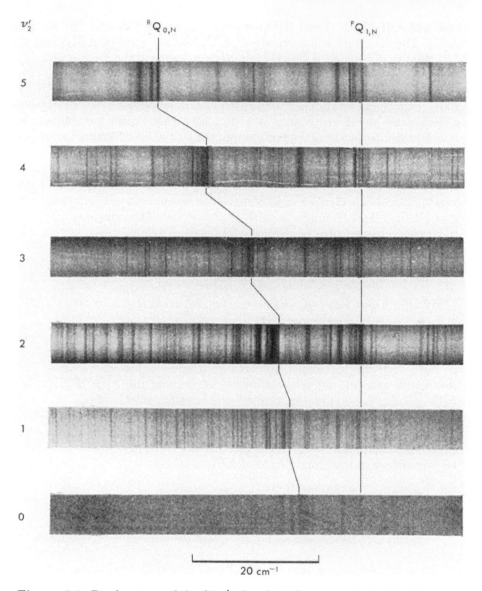

Figure 6.5: Band centres of the $(0, v_2', 0) - (0, 0, 0)$ progression in the absorption spectrum of AsH_2. From Dixon et al., *Proc. Roy. Soc. A*, 305:271, 1968 [16]. Reproduced with permission from The Royal Society of Chemistry.

The bands analysed were $v_2' = 1$ to $v_2' = 3$. The $v_2' = 0$ is too weak, while the higher bands are too diffuse for rotational analysis, indicating rapidly increasing predissociaton. Above $v_2' = 5$, no absorption spectra could be recorded. AsH_2 may either dissociate to As + 2H, As + H_2, or AsH + H, of which the latter is the most probable process. Combination of the ground states of AsH $(X\,^3\Sigma^-)$ and H(^2S) leads to the states $^2B_1$ and $^4B_1$ for bent AsH$_2$. The $^2B_1$ state will correlate with the ground state of AsH$_2$, and the $^4B_1$ state may give rise to a forbidden predissociation of the $^2A_1$ excited state through strong spin-orbit coupling.

Spin-orbit coupling increases considerably when passing from NH$_2$ to AsH$_2$, and, hence, rotational structure changes from a Hund's case (b) towards a Hund's case (a) characteristics. N is no longer a good quantum number, but may be retained as a formal quantum number for the designation of levels and the sorting of branches.

If the spin-rotation interaction constants are small compared with the rotational constants, the rotational levels confirm to Hund's case (b). The Hamiltonian matrix can then be approximately separated into F_1 matrices, with $J = N + 1/2$, and F_2 matrices, with $J = N - 1/2$. The rotational levels are then related to the corresponding singlet levels as follows:

$$F_1(N, K_a, K_c) = E_r(N, K_a, K_c) + \frac{1}{2(N+1)}\left(\epsilon_{aa}N_a^2 + \epsilon_{bb}N_b^2 + \epsilon_{cc}N_c^2\right) \quad (6.2)$$

$$F_2(N, K_a, K_c) = E_r(N, K_a, K_c) - \frac{1}{2N}\left(\epsilon_{aa}N_a^2 + \epsilon_{bb}N_b^2 + \epsilon_{cc}N_c^2\right). \quad (6.3)$$

For large spin-rotation coupling the rotation cannot be separated from the spin coupling and, hence, the full interaction matrix must be diagonalized. In AsH$_2$ the principal departure between the equations above arises from the matrix elements between the basis functions of the $F_1(N, K_a, K_c)$ and $F_2(N+1, K_a, K_c+1)$ blocks, through the large magnitude of ϵ_{aa}.

The effect of these interactions is indicated in Figure 6.6, and is also apparent in the observed spectra in Figure 6.7. It should be noted that such an effect would not occur if ϵ_{aa} were negative, since the spin doublets would then be inverted, as in the ground state of AsH$_2$.

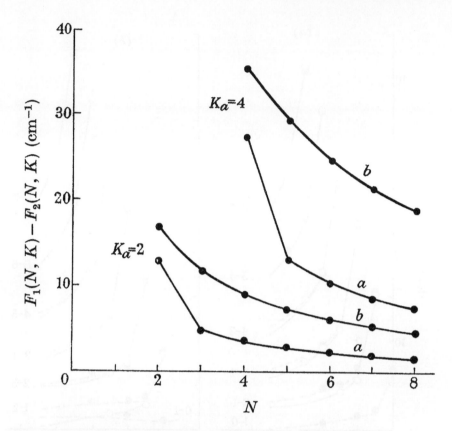

Figure 6.6: Doublet splittings calculated for a symmetric top for $\epsilon_{aa} = 10\,\mathrm{cm}^{-1}$ and $\bar{B} = 4\,\mathrm{cm}^{-1}$, a, including the interaction between the levels $F_1(N, K_a, J)$ and $F_2(N+1, K_a, J)$, and b, neglecting this interaction, i.e., Hund's case (b) coupling. Reproduced by permission from R. N. Dixon et al., Proc. Roy. Soc. A305:271, 1968 [16].

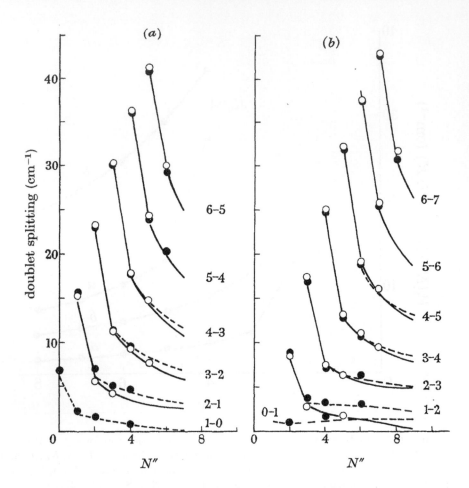

Figure 6.7: A comparison between the doublet splittings observed in some of the branches of the $(0,3,0)–(0,0,0)$ band of AsH_2 and those calculated with the fitted parameters: (a) ^RR branches; (b) ^PP branches. The branches are labelled by the values of $K_a' - K_a''$, and the asymmetry components by full circles $(K_a + K_c = N)$ and open circles $(K_a + K_c = N+1)$. From Dixon et al., *Proc. Roy. Soc. A*, 305:271, 1968 [16]. Reproduced with permission from The Royal Society of Chemistry.

The calculations for Figure 6.6 indicated that for the higher values of K_a, the F_1 levels with $N = K_a$ should be mixed with the F_2 levels with $N = K_a+1$ to a sufficient extent that satellite transitions should occur. These violate the ΔN selection rules but satisfy the ΔJ selection rules (tendency towards Hund's case (a)). We have found several such satellite lines in the $3\nu_2'$ band, and have verified their assignments through combination differences (see Table 6.14 at the end of this chapter). The identification of the F_1 and F_2 series of levels is, therefore, unambiguous.

The molecule is extremely nonrigid in the excited state, and we have found it necessary to extend the Hamiltonian

$$H = AN_a^2 + BN_b^2 + CN_c^2 + \frac{1}{4} \sum_{\alpha,\beta,\gamma,\delta} \tau_{\alpha,\beta,\gamma,\delta} N_\alpha N_\beta N_\gamma N_\delta + \sum_{\alpha,\beta} \epsilon_{\alpha,\beta} S_\alpha N_\beta, \quad (6.4)$$

where $\alpha = a, b, c$, etc., by the addition of the terms $H_K N_a^6 + \eta_{aaaa} N_a^3 S_a$. For a discussion of strong spin-rotation in a non-rigid molecule, see Dixon and Duxbury [15].

6.2.3   Effects of Spin-Orbit Coupling on the Spin-Rotation Interaction in the AsH_2 Radical

G. Duxbury and A. Alijah, International Symposium on Molecular Spectroscopy 2014 [21]

The occurence of predissociation in the electronic spectrum of AsH_2 is very dependent upon the magnitude of the spin-orbit coupling parameter of the central atom. Making use of Table 5.6 in *The Spectra and Dynamics of Diatomic Molecules* by Hélène Lefebvre-Brion and Robert W. Field [22], it is possible to appreciate the rapid rate of increase of the spin-orbit constants associated with the heavy central atom in the di-hydrides NH_2, PH_2 and AsH_2. The spin-orbit constants range from 42.7 cm^{-1} for NH_2, to 191.3 cm^{-1} for PH_2, and 1178 cm^{-1} for AsH_2. The effects of spin-orbit coupling may be seen in Figure 6.5 via the change in separation of the central ${}^rQ_{0,N}$ and ${}^PQ_{1,N}$ subbands as the value of v_2' increases from 0 to 5. As the value of v_2' increases beyond 2, the spectrum becomes more and more fuzzy as the effects of predissociation become more obvious. This means that unlike the example of the behaviour of PH_2, where the vibronic level pattern can be followed below and above the barrier to linearity, in AsH_2 and AsD_2, the absorption spectrum becomes completely diffuse below the barrier to linearity in the $\tilde{A}\,^2A_1$ state. Predissociation can also be seen in the absorption spectra obtained by cavity ringdown spectroscopy by Zhao and Chen, which are shown in Figures 6.8 to 6.12.

The change in the magnitude of the doublet splittings as v_2' increases may be seen in the plots of the doublet splittings showing the spin-uncoupling as a result of the increase of overall rotation. In the absorption spectrum of SbH_2, recorded in 1967 by T. Barrow in the Chemistry Department at Sheffield University (T. Barrow, PhD thesis), all the absorption features showed the effects of predissociation, consistent with a spin-orbit constant of 2834 cm^{-1} for the central atom of SbH_2.

6.2.4   Diagrams for AsH_2

Dongfeng Zhao (PhD thesis, University of Science and Technology of China, Hefei) [23] and Yang Chen.

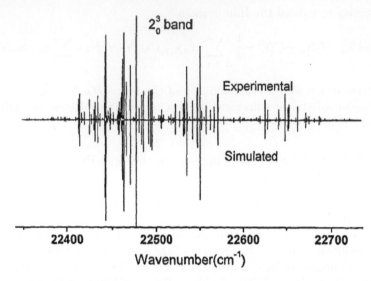

Figure 6.8: Start of predissociation. Reproduced with permission from the author.

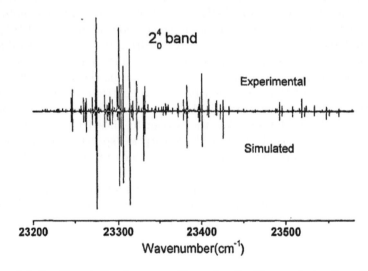

Figure 6.9: Predissociation increases. Reproduced with permission from the author.

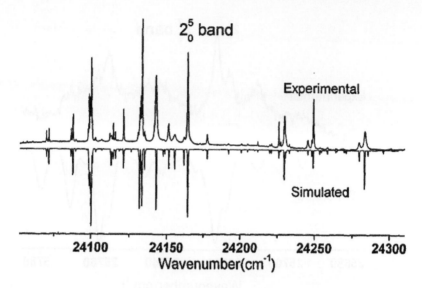

Figure 6.10: Predissociation visible. Reproduced with permission from the author.

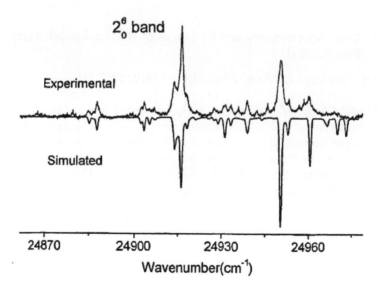

Figure 6.11: Predissociation effects large. Reproduced with permission from the author.

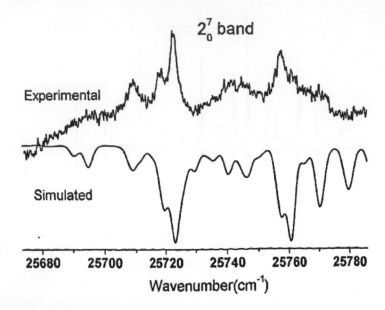

Figure 6.12: Predissociation effects very large, last band visible. Reproduced with permission from the author.

6.2.5 Laser Spectroscopy and Dynamics of the Jet-Cooled AsH_2 Free Radical

He and Clouthier, *J. Chem. Phys.*, 125:154312, 2007 [24].

He and Clouthier studied the AsH_2 jet-cooled free radical using laser in-duced fluorescence (LIF). This enabled them to obtain sharp spectra from the bands at the bottom of the potential well, shown in Figures 6.13 to 6.15, and the transitions in Table 6.7. They also used laser excitation of the jet-cooled AsH_2 to obtain the low-resolution spectrum shown in Figure 6.16. Figure 6.17 shows the effect of collisional quenching on the lifetimes for different states at the bottom of the potential well.

Table 6.7: Key Rotational Transitions and Vibrational Intervals (In cm^{-1}) Identified In the LIF Spectrum of AsH$_2$

Band	Rotational transition				Band origin	$G_0(\nu')$	ΔG
	$0_{00} - 1_{10}(F_2)$	$0_{00} - 1_{10}(F_1)$	$1_{10} - 0_{00}(F_2)$	$1_{10} - 0_{00}(F_1)$			
0_0^0	19893.98	19895.11	19929.17	19932.45	19909.5	0	$\cdots$
2_0^1	20744.5	20745.55	20781.10	20785.36	19909.5	850.43	850.43
2_0^2	21589.87	21590.98	21628.48	21633.86	21605.3	1695.86	845.43
2_0^3	22429.43	22430.53	22470.27	22477.25	22444.9	2535.86	839.55
2_0^4	23262.34	23263.45	23305.85	23313.84	23277.9	3368.33	832.92
1_0^1	22006.82	22007.83	22041.68	22044.91	22022.2	2112.71	$\cdots$
1_0^0	222842.60	22843.79	22878.53	22882.66	22858.1	2948.67	835.96
2_1^2				19799.77	19803.97		

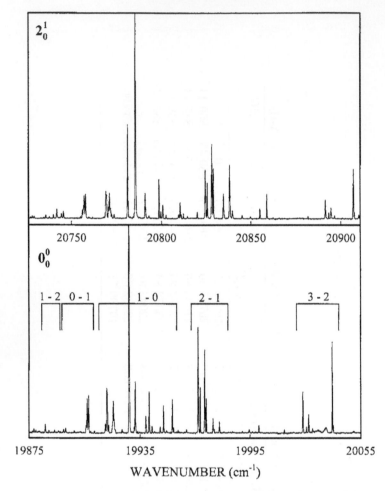

Figure 6.13: LIF spectra of the 2_0^1 (top panel) and 0_0^0 (bottom panel) bands of jet-cooled AsH$_2$. The spectra have been aligned vertically so that the $1_{10}(F_2) - 0_{00}(F_1)$ rovibronic transitions occur in the same position. In each case, the strongest feature is off scale. In the 0_0^0 band spectrum, the vertical leaders denote the extent of the individual subbands labeled as $K_a' - K_a''$. The subbands are further apart in the 2_0^1 spectrum due to an increase in the A rotational constant on excitation of the bending mode. Reproduced from He and Clouthier, *J. Chem. Phys.*, 126:154312, 2007 [24], with the permission of AIP Publishing.

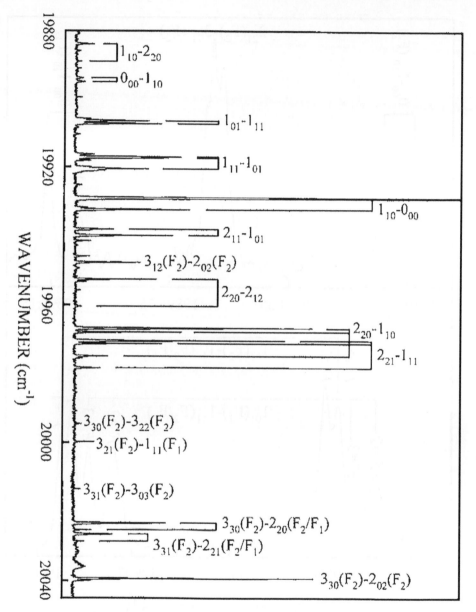

Figure 6.14: The central portion of the 0_0^0 band LIF spectrum of jet-cooled AsH$_2$ showing some of the rotational assignments, including asymmetry and spin splittings. Reproduced from He and Clouthier, *J. Chem. Phys.*, 126:154312, 2007 [24], with the permission of AIP Publishing.

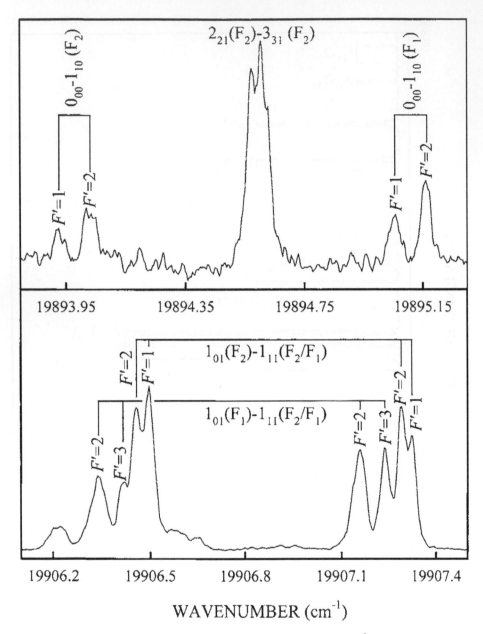

Figure 6.15: High-resolution LIF spectra of portions of the 0_0^0 band of jet-cooled AsH$_2$. The top panel shows the large spin splitting of the $0_{00} - 1_{10}$ rotational transition and the much smaller hyperfine splitting of each spin component. The central feature shows the $2_{21} - 3_{31}(F_2)$ rotational line, which exhibits partially resolved hyperfine splittings. The bottom panel shows the complex pattern of partially resolved hyperfine structure observed for the $1_{01} - 1_{11}$ transition. Reproduced from He and Clouthier, *J. Chem. Phys.*, 126:154312, 2007 [24], with the permission of AIP Publishing.

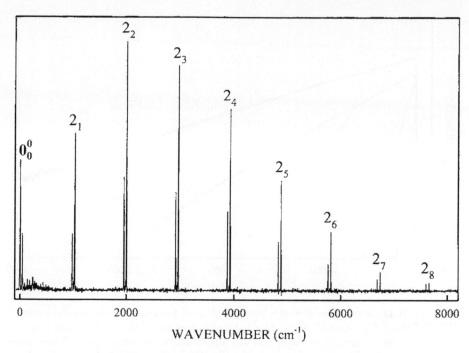

Figure 6.16: The low-resolution emission spectrum of jet-cooled AsH_2 obtained on laser excitation of the $1_{10} - 0_{00}(F_2 - F_1)$ rotational transition of the 0_0^0 band. The wave number scale is displacement from the excitation laser frequency, giving a direct measure of the ground state vibrational energy. The resonance fluorescence at 0.0 cm^{-1} is enhanced by scattered laser light. The lower-state vibronic assignments are given, and each vibronic band consists of two rotational lines, a weaker transition down to the 0_{00} level, and a stronger transition down to 2_{20}. Reproduced from He and Clouthier, *J. Chem. Phys.*, 126:154312, 2007 [24], with the permission of AIP Publishing.

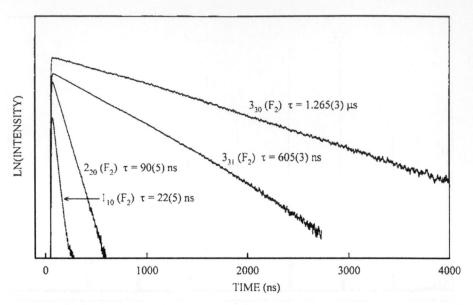

Figure 6.17: Examples of single rotational level fluorescence decays from the 0^0 level of the $\tilde{A}\,^2A_1$ state of AsH$_2$. The natural logarithm of the fluorescence intensity is plotted as a function of time, and the upper-state rotational level, fluorescence lifetime, and standard error are given for each decay curve. The longer lifetime decay curves are affected by collisional quenching. Reproduced from He and Clouthier, *J. Chem. Phys.*, 126:154312, 2007 [24], with the permission of AIP Publishing.

6.2.6 Absorption Spectra of AsH$_2$ Radical In $435-510$ nm by Cavity Ringdown Spectroscopy

D. Zhao, C. Qin, M. Ji, Q. Zhang, and Y. Chen, *J. Mol. Spectrosc.*, 256:192, 2009 [25].

Zhao et al. studied the radical using cavity ringdown spectroscopy (CRD), which enabled them to work at the bottom of the potential well. The schematic diagram of their experimental setup is shown in Figure 6.18, and the resulting spectra in Figure 6.19, with more detail shown in Figure 6.20.

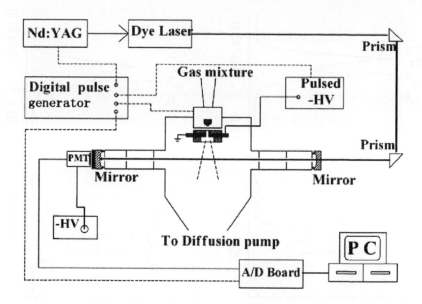

Figure 6.18: Schematic diagram of the pulsed discharge and the CRD apparatus. Reprinted from Zhao et al., *J. Mol. Spectrosc.*, 256:192, 2009 [25]. Copyright (2009), with permission from Elsevier.

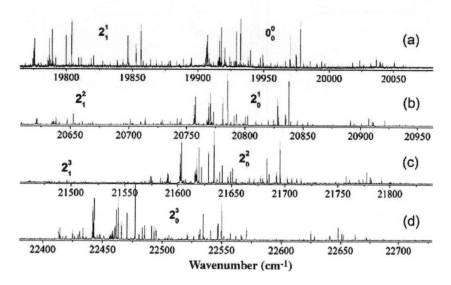

Figure 6.19: CRD spectra of AsH_2 in 435–510 nm. Seven vibronic bands were assigned to two progressions of 0_0^0, 2_0^n, 2_1^n, $n = 1 - 3$. Reprinted from Zhao et al., *J. Mol. Spectrosc.*, 256:192, 2009 [25]. Copyright (2009), with permission from Elsevier.

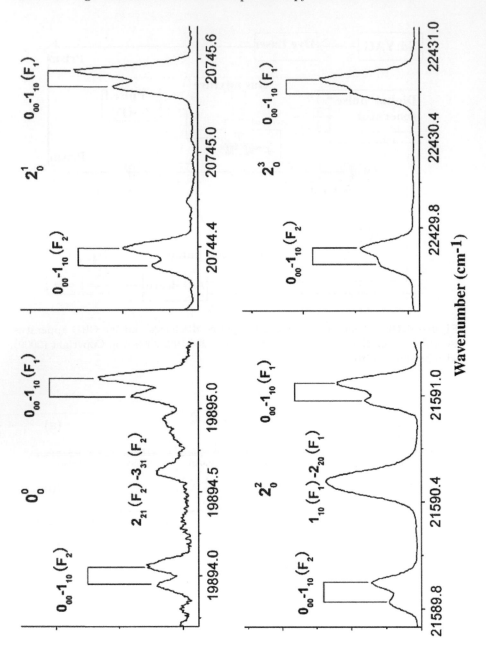

Figure 6.20: Portions of the CRD spectra of AsH₂. The four panels show the large spin splittings of the $0_{00} - 1_{10}$ rotational transitions, and the much smaller hyperfine splittings for each spin components in the 0_0^0 and 2_0^n, $n = 1 - 3$ bands, respectively. Reprinted from Zhao et al., *J. Mol. Spectrosc.*, 256:192, 2009 [25]. Copyright (2009), with permission from Elsevier.

6.2.7 Toward an Improved Understanding of the AsH$_2$ Free Radical: Laser Spectroscopy, Ab Initio Calculations, and Normal Coordinate Analysis

Grimminger and Clouthier, *J. Chem. Phys.*, 137:224307, 2012 [26].

Spectra of the $\tilde{A}\,^2A_1 - \tilde{X}\,^2B_1$ transition of the jet-cooled AsD$_2$ and AsHD isotopologues of the arsino radical have been studied by LIF and wavelength-resolved emission techniques; see Figures 6.21 and 6.22.

A high-resolution spectrum of the AsD$_2$ 0_0^0 band has been recorded, and an improved r_0 structure [$r'_0 = 1.487(4)$ Å, $\theta'_0 = 123.0(2)°$] for the $\tilde{A}$ state has been determined from the rotational constants. To aid in the analysis of the vibrational levels, an ab initio potential energy surface of the $\tilde{X}\,^2B_1$ state has been constructed, and the rovibronic energy levels of states on that potential have been determined using a variational method. The vibrational levels observed in wavelength-resolved emission spectra have been fitted to a

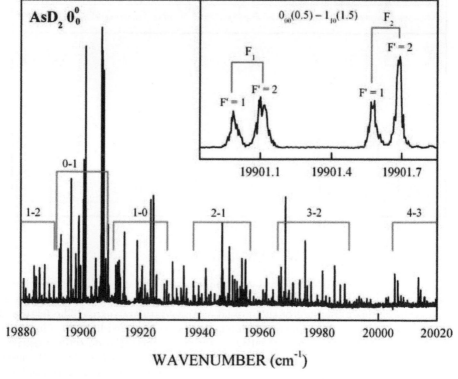

Figure 6.21: High-resolution LIF spectrum of the AsD$_2$ 0_0^0 transition. The approximate positions of the subbands have been labeled with leaders. The inset shows the large spin splitting for the $0_{00} - 1_{10}$ rotational transition as well as the smaller hyperfine splitting. Reproduced from Grimminger and Clouthier, *J. Chem. Phys.*, 137:224307, 2012 [26], with the permission of AIP Publishing.

local mode Hamiltonian with most anharmonic parameters fixed at ab initio values, and the resulting harmonic frequencies have been used to perform a normal coordinate analysis, which yielded an improved set of quadratic force constants and an estimate of the equilibrium ground state structure. The transitions, molecular parameters, and energy levels are shown in Tables 6.8 to 6.10.

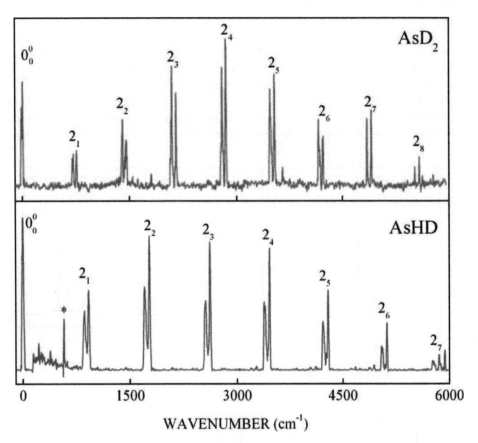

Figure 6.22: Single vibronic level emission spectra. The upper trace shows AsD$_2$, and the lower shows AsHD. The horizontal axis is displacement from the excitation frequency, giving a measure of the ground state vibrational energy. The feature at 0.0 cm^{-1} displacement has increased intensity due to scattered light. The increased baseline at $\sim$ 150 cm^{-1} in the AsHD spectrum is where the detector sensitivity was increased, and is due to scattered broadband dye fluorescence from the laser. The asterisk marks an impurity feature not due to AsHD fluorescence. Reproduced from Grimminger and Clouthier, *J. Chem. Phys.*, 137:224307, 2012 [26], with the permission of AIP Publishing.

Table 6.8: Observed Transitions of the $\tilde{A}\,^2A_1 - \tilde{X}\,^2B_1$ 0_0^0 Band of Jet Cooled AsD$_2$

N'	K_a'	K_c'	J'	F'	N''	K_a''	K_c''	J''	F''	Observed	Obs.–Calc.
0	0	0	0.5	1	1	1	0	0.5	2	19900.9915	−0.0038
0	0	0	0.5	1	1	1	0	0.5	1	19901.0129	−0.0083
0	0	0	0.5	1	1	1	0	1.5	2	19901.5733	−0.0065
0	0	0	0.5	1	1	1	0	1.5	1	19901.5733	−0.0058
0	0	0	0.5	1	1	1	0	1.5	0	19901.5733	−0.0058
0	0	0	0.5	2	1	1	0	0.5	2	19901.0981	−0.0107
0	0	0	0.5	2	1	1	0	0.5	1	19901.1172	
0	0	0	0.5	2	1	1	0	1.5	3	19901.6862	−0.0097
0	0	0	0.5	2	1	1	0	1.5	2	19901.6862	−0.0071
0	0	0	0.5	2	1	1	0	1.5	1	19901.6862	−0.0064
1	0	1	0.5	1	2	1	1	1.5	2	19892.9551	0.003
1	0	1	0.5	1	2	1	1	1.5	1	19892.9551	−0.0017
1	0	1	0.5	1	2	1	1	1.5	0	19892.9551	−0.0046
1	0	1	0.5	1	2	1	1	2.5	2	19893.5151	
1	0	1	0.5	1	2	1	1	2.5	1	19893.5151	
1	0	1	0.5	2	1	1	1	0.5	2	19907.3659	
1	0	1	0.5	2	2	1	1	1.5	3	19892.9287	−0.0053
1	0	1	0.5	2	2	1	1	1.5	2	19892.9287	−0.01
1	0	1	0.5	2	2	1	1	2.5	3	19893.5151	
1	0	1	0.5	2	2	1	1	2.5	2	19893.5151	
1	1	1	0.5	1	1	0	1	1.5	1	19913.0696	
1	1	1	0.5	1	1	0	1	1.5	0	19913.0696	
1	1	1	0.5	1	2	2	1	1.5	2	19896.8740	
1	1	1	0.5	2	1	0	1	0.5	2	19912.9070	0.0053
1	1	1	0.5	2	1	0	1	0.5	1	19912.9070	
1	1	1	0.5	2	1	0	1	1.5	3	19913.0512	0.0116
1	1	1	0.5	2	1	0	1	1.5	2	19913.0512	0.008
1	1	1	0.5	2	2	2	1	1.5	3	19896.8575	0.0054
1	1	0	0.5	1	0	0	0	0.5	2	19918.9605	0.005
1	1	0	0.5	1	2	2	0	1.5	1	19895.9063	
1	1	0	0.5	1	2	2	0	1.5	0	19895.9459	0.0014
1	1	0	0.5	2	0	0	0	0.5	2	19918.9232	0.0064
1	1	0	0.5	2	0	0	0	0.5	1	19918.9232	0.0024
1	1	0	0.5	2	2	2	0	1.5	3	19895.8894	0.0059
1	1	0	0.5	2	2	2	0	1.5	2	19895.8571	
1	0	1	1.5	0	2	1	1	1.5	1	19892.8079	0.0075
1	0	1	1.5	1	2	1	1	1.5	2	19892.8079	0.0063
1	0	1	1.5	1	2	1	1	2.5	2	19893.3776	0.0081
1	0	1	1.5	1	2	1	1	2.5	1	19893.3776	0.0074
1	0	1	1.5	2	2	1	1	1.5	3	19892.8079	−0.0025

Table 6.8: Observed Transitions of the $\tilde{A}\,^2A_1 - \tilde{X}\,^2B_1\ 0^0_0$ Band of Jet Cooled AsD$_2$

N'	K'_a	K'_c	J'	F'	N''	K''_a	K''_c	J''	F''	Observed	Obs.–Calc.
1	0	1	1.5	3	1	1	1	1.5	3	19907.7316	−0.0028
1	0	1	1.5	3	2	1	1	2.5	4	19893.4729	−0.0062
1	0	1	1.5	3	2	1	1	2.5	3	19893.4729	−0.0043
1	1	1	1.5	3	1	0	1	1.5	3	19914.6380	0.0046
1	1	1	1.5	3	1	0	1	1.5	2	19914.6380	0.0011
1	1	1	1.5	3	2	2	1	2.5	4	19899.4953	0.0119
1	1	1	1.5	3	2	2	1	2.5	3	19899.4953	0.0124
1	1	0	1.5	2	0	0	0	0.5	2	19920.5470	−0.0043
1	1	0	1.5	2	0	0	0	0.5	1	19920.5470	−0.0082
1	1	0	1.5	2	2	2	0	2.5	3	19898.5134	−0.0029
1	1	0	1.5	3	0	0	0	0.5	2	19920.6003	−0.0038
1	1	0	1.5	3	2	2	0	1.5	3	19897.5748	0.004
1	1	0	1.5	3	2	2	0	2.5	4	19898.5801	0.0072
2	0	2	1.5	2	3	1	2	2.5	3	19884.5607	0.001
2	0	2	1.5	3	2	1	2	1.5	3	19907.0202	0.0009
2	0	2	1.5	3	3	1	2	2.5	3	19884.5405	0.0004
2	1	2	1.5	2	3	2	2	2.5	3	19889.6270	0.0072
2	1	2	1.5	3	3	2.	2	2.5	4	19889.5922	0.0038
2	1	1	1.5	3	1	0	1	0.5	2	19923.4086	0.0092
2	1	1	1.5	3	1	0	1	1.5	3	19923.5393	0.0021
2	1	1	1.5	3	3	2	1	2.5	4	19886.3564	0.0056
2	2	1	1.5	3	1	1	1	0.5	2	19939.7897	−0.0075
2	2	1	1.5	3	1	1	1	1.5	3	19940.2087	−0.0076
2	2	0	1.5	3	1	1	0	0.5	2	19937.9700	−0.0044
2	2	0	1.5	3	1	1	0	1.5	3	19938.5655	0.004
2	2	0	1.5	3	2	1	2	1.5	3	19930.8180	0.0026
2	2	0	1.5	3	3	3	0	2.5	4	19900.0456	0.0014
2	2	0	1.5	3	3	3	0	2.5	3	19900.0456	−0.0059
2	0	2	2.5	3	2	1	2	1.5	3	19906.8908	−0.0012
2	0	2	2.5	3	3	1	2	2.5	4	19884.4111	0.0025
2	0	2	2.5	4	1	1	0	1.5	3	19914.7240	0.0077
2	0	2	2.5	4	2	1	2	2.5	4	19907.2849	0.0085
2	0	2	2.5	4	3	1	$\tilde{A}$	3.5	5	19885.2728	−0.0018
2	0	2	2.5	4	3	1	2	3.5	4	19885.2728	−0.0027
2	1	2	2.5	3	3	2	2	3.5	4	19891.1069	0.0163
2	1	2	2.5	4	2	0	2	2.5	4	19913.2957	0.0101
2	1	2	2.5	4	3	2	2	3.5	5	19891.1407	0.0115
2	1	2	2.5	4	3	2	2	3.5	4	19891.1407	
2	1	1	2.5	1	2	2	1	2.5	1	19909.1309	0.015
2	1	1	2.5	1	3	2	1	3.5	2	19887.9633	
2	1	1	2.5	2	2	2	1	2.5	2	19909.1445	0.0037

Table 6.8: Observed Transitions of the $\tilde{A}\,^2A_1 - \tilde{X}\,^2B_1$ 0_0^0 Band of Jet Cooled AsD_2

N'	K_a'	K_c'	J'	F'	N''	K_a''	K_c''	J''	F''	Observed	Obs.–Calc.
2	1	1	2.5	2	3	2	1	3.5	3	19888.0216	−0.0091
2	1	1	2.5	3	2	2	1	2.5	3	19909.1821	0.007
2	1	1	2.5	3	3	2	1	3.5	4	19888.0634	−0.0022
2	1	1	2.5	3	3	2	1	3.5	3	19888.6634	−0.0009
2	1	1	2.5	4	1	0	1	1.5	3	19924.3734	0.0071
2	1	1	2.5	4	2	2	1	2.5	4	19909.2304	0.0141
2	1	1	2.5	4	3	2	1	2.5	4	19887.1766	−0.0032
2	1	1	2.5	4	3	2	1	3.5	5	19888.1120	0.0021
2	1	1	2.5	4	3	2	1	3.5	4	19888.1120	0.0057
2	2	1	2.5	3	3	3	1	3.5	4	19906.2301	−0.0016
2	2	1	2.5	4	1	1	1	1.5	3	19943.8087	−0.0083
2	2	1	2.5	4	2	1	1	2.5	4	19929.5505	−0.0112
2	2	1	2.5	4	3	3	1	3.5	4	19906.2770	0.0028
2	2	0	2.5	3	3	3	0	3.5	4	19905.0556	−0.0015
2	2	0	2.5	4	1	1	0	1.5	3	19942.1544	−0.0047
2	2	0	2.5	4	2	1	2	2.5	4	19934.7245	0.0052
2	2	0	2.5	4	3	3	0	3.5	5	19905.1001	−0.0033
3	1	3	2.5	4	4	2	3	3.5	5	19881.6713	−0.0125
3	1	3	2.5	4	4	2	3	3.5	4	19881.6713	−0.0091
3	1	2	2.5	4	2	0	2	1.5	3	19928.0417	−0.0016
3	1	2	2.5	4	2	2	0	1.5	3	19919.9630	−0.0018
3	1	2	2.5	4	3	2	2	2.5	3	19905.2562	0.0039
3	2	1	2.5	4	1	1	1	1.5	3	19954.5687	−0.0033
3	2	1	2.5	4	3	1	3	2.5	4	19932.2256	−0.0002
3	3	1	2.5	3	3	0	3	2.5	3	19962.2417	−0.0001
3	3	1	2.5	4	2	2	1	1.5	3	19968.5643	−0.0094
3	3	1	2.5	4	2	2	1	2.5	4	19969.5879	
3	3	1	2.5	4	3	0	3	2.5	4	19962.2214	0.0003
3	3	1	2.5	4	3	2	1	2.5	4	19947.5722	−0.0026
3	3	0	2.5	4	2	2	0	1.5	3	19967.0983	−0.0118
3	3	0	2.5	4	3	2	2	2.5	4	19952.3981	−0.0003
3	0	3	3.5	5	2	1	1	2.5	4	19915.0751	0.0016
3	1	3	3.5	5	2	2	1	2.5	4	19919.5253	−0.0135
3	1	3	3.5	5	4	2	3	4.5	6	19882.9270	−0.0059
3	1	3	3.5	5	4	2	3	4.5	5	19882.9270	−0.009
3	1	2	3.5	4	2	2	0	2.5	3	19921.5251	−0.0046
3	1	2	3.5	4	3	2	2	3.5	4	19906.7110	−0.0117
3	1	2	3.5	5	2	0	2	2.5	4	19928.9127	0.0001
3	1	2	3.5	5	2	2	0	2.5	4	19921.5728	0.0046
3	1	2	3.5	5	3	2	2	3.5	5	19906.7520	−0.0042
3	2	2	3.5	5	2	1	2	2.5	4	19947.4175	0.0067

Table 6.8: Observed Transitions of the $\tilde{A}^2A_1 - \tilde{X}^2B_1$ 0_0^0 Band of Jet Cooled AsD_2

N'	K_a'	K_c'	J'	F'	N''	K_a''	K_c''	J''	F''	Observed	Obs.–Calc.
3	2	2	3.5	5	3	1	2	3.5	4	19925.4139	0.0041
3	2	2	3.5	5	4	3	2	4.5	6	19895.6471	0.0082
3	2	1	3.5	5	2	1	1	2.5	4	19942.4647	0.0019
3	3	1	3.5	5	2	2	1	2.5	4	19975.2419	0.0028
3	3	1	3.5	5	3	0	3	3.5	5	19968.1331	−0.0163
3	3	1	3.5	5	3	2	1	3.5	5	19954.1110	
4	1	3	3.5	5	3	0	3	2.5	4	19933.6660	−0.0048
4	1	3	3.5	5	3	2	1	2.5	4	19919.0233	−0.0011
4	1	3	3.5	5	4	2	3	3.5	5	19903.5304	−0.0138
4	2	3	3.5	5	2	1	1	2.5	4	19957.9035	
4	2	3	3.5	5	3	1	3	2.5	4	19949.8526	0.0004
4	2	2	3.5	3	3	3	0	2.5	2	19932.5852	−0.0045
4	2	2	3.5	4	3	3	0	2.5	3	19932.5525	−0.0059
4	2	2	3.5	5	3	1	2	2.5	4	19940.8213	0.0077
4	2	2	3.5	5	3	3	0	2.5	4	19932.5164	−0.0045
4	3	2	3.5	5	3	2	2	2.5	4	19971.2018	−0.0153
4	3	2	3.5	5	4	0	4	3.5	5	19964.4277	0.0068
4	3	1	3.5	4	2	2	1	2.5	3	19988.4777	0.007
4	3	1	3.5	4	3	0	3	2.5	3	19981.0648	−0.0045
4	3	1	3.5	5	2	2	1	2.5	4	19988.4574	0.0155
4	3	1	3.5	5	3	2	1	2.5	4	19966.4074	0.002
4	3	1	3.5	5	4	2	3	3.5	4	19950.9225	0.0007
4	4	1	3.5	5	3	1	3	2.5	4	20023.3821	−0.0099
4	4	1	3.5	5	3	3	1	2.5	4	20006.6183	−0.0062
4	4	0	3.5	4	3	3	0	2.5	3	20005.5595	−0.0095
4	4	0	3.5	4	4	3	2	3.5	4	19983.4259	
4	4	0	3.5	5	3	1	2	2.5	4	20013.8365	0.0135
4	4	0	3.5	5	3	3	0	2.5	4	20005.5297	−0.0006
4	0	4	4.5	6	3	1	2	3.5	5	19915.2585	−0.0021
4	1	3	4.5	6	3	0	3	3.5	5	19934.4492	−0.0066
4	1	3	4.5	6	4	2	3	4.5	6	19904.9377	−0.0019
4	2	2	4.5	5	3	3	0	3.5	4	19935.5266	0.0051
4	2	2	4.5	6	3	1	2	3.5	5	19943.1816	0.0069
4	2	2	4.5	6	4	1	4	4.5	6	19935.6526	−0.0028
4	3	2	4.5	6	3	2	2	3.5	5	19975.9244	0.0138
4	3	1	4.5	3	3	2	1	3.5	2	19971.0222	−0.0091
4	3	1	4.5	5	3	0	3	3.5	4	19985.1195	0.0035
4	3	1	4.5	6	3	0	3	3.5	5	19985.1483	0.0051
4	3	1	4.5	6	4	2	3	4.5	6	19955.6233	−0.0037
4	4	0	4.5	6	3	1	2	3.5	5	20022.2377	−0.0136
4	4	0	4.5	6	3	3	0	3.5	5	20014.6298	−0.0075

Table 6.8: Observed Transitions of the $\tilde{A}\,^2A_1 - \tilde{X}\,^2B_1$ 0_0^0 Band of Jet Cooled AsD$_2$

N'	K'_a	K'_c	J'	F'	N''	K''_a	K''_c	J''	F''	Observed	Obs.–Calc.
4	4	0	4.5	6	4	3	2	4.5	6	19992.4895	0.0082
5	3	3	4.5	3	4	2	3	3.5	2	19973.4215	
5	3	2	4.5	6	4	2	2	3.5	5	19964.2698	0.0166
5	4	2	4.5	6	5	1	4	4.5	6	19992.5545	0.0183
5	4	2	4.5	6	5	3	2	4.5	6	19970.5078	0.005
5	2	4	5.5	7	4	1	4	4.5	6	19956.7880	−0.0078
5	3	3	5.5	7	4	2	3	4.5	6	19977.2531	
5	3	3	5.5	7	5	0	5	5.5	7	19969.8413	0
5	4	2	5.5	7	4	3	2	4.5	6	20013.7433	
5	4	1	5.5	7	4	1	3	4.5	6	20023.0699	0.0034
5	4	1	5.5	7	5	3	3	5.5	7	19986.1477	
6	2	5	5.5	7	5	1	5	4.5	6	19961.0406	0.0001
6	2	5	6.5	8	5	1	5	5.5	7	19961.7542	

Transitions without an "Obs. – Calc." entry were overlapped or otherwise compromised, so were not used in the least squares analysis.

Table 6.9: Molecular Parameters for AsD_2 (In cm^{-1}) Determined from Fitting the Rotationally Resolved 0_0^0 Band

Parameter	AsD_2 $\tilde{A}(0^0)^a$	AsD_2 $\tilde{X}(0_0)^b$	AsH_2 $\tilde{A}(0^0)^c$
T_0	19908.9252(11)	0.0	19909.4531
A	8.7310(5)	3.8814852	17.2065
B	2.4687(2)	3.5866129	4.91956
C	$1.8954_1(3)$	1.442100	3.74042
Δ_N		1.165×10^{-4}	
Δ_{NK}		-3.8407×10^{-4}	-0.600×10^{-3}
Δ_K	$6.04_{39} \times 10^{-3}$	5.7150×10^{-4}	0.25623×10^{-1}
δ_N		5.3010×10^{-5}	
δ_K		-3.446×10^{-5}	
ϵ_{aa}	$2.329_6(3)$	-5.67439×10^{-1}	4.7179
ϵ_{bb}	0.0443(15)	-1.95983×10^{-1}	0.0826
ϵ_{cc}	$-0.1092(19)$	1.935×10^{-3}	-0.2028
Δ_N^s		2.12×10^{-5}	
$\Delta_{NK}^s + \Delta_{KN}^s$		-5.417×10^{-5}	
Δ_K^s	$-5.1_6(2) \times 10^{-3}$	1.954×10^{-4}	-0.02369
δ_N^s		1.28×10^{-5}	
$a_F(As)$	0.0567(11)	1.99×10^{-3}	0.0511
$T_{aa}(As)$		-9.5896×10^{-3}	-0.0087
$T_{bb}(As)$		1.0736×10^{-2}	0.0168
$\chi_{aa}(As)$		1.054×10^{-3}	
$\chi_{bb}(As)$		-4.7713×10^{-3}	
$C_{aa}(As)$		1.08×10^{-5}	
$C_{bb}(As)$		6.74×10^{-6}	
$C_{cc}(As)$		9.0×10^{-7}	

[a]This work. The numbers in parenthesis are 1σ error limits and are right justified to the last digit on the line. Numbers below the line are added to reproduce the original data to full accuracy.
[b]AsD_2 ground state values from Reference [18].
[c]AsH_2 excited state values from Reference [24].

Table 6.10: Measured Ground State Vibrational Energy Levels (In cm^{-1}) of AsH_2 Isotopologues

Level	AsH_2 Energy[a]	Obs.–Calc.	AsD_2 Energy	Obs.–Calc.	AsHD Energy	Obs.–Calc.
2_1	981.4	−1.4	704.7	0.6	855.7	−0.7
2_2	1953.0	−3.5	1403.9	−0.3	1704.3	−0.7
2_3	2918.7	−2.5	2096.7	−3.4	2547.4	0.0
2_4	3875.7	−1.1	2795.0	2.9	3384.7	−0.4
2_5	4824.8	1.3	3478.4	−1.6	4214.8	0.6
2_6	5763.1	2.0	4165.6	1.8	5039.9	0.1
2_7	6693.1	3.5	4842.9	−0.6	6670.2	1.5
2_8	7609.5	0.3	$\cdots$	$\cdots$	$\cdots$	$\cdots$
2_9	8516.8	−2.9	$\cdots$	$\cdots$	$\cdots$	$\cdots$
1_1	2095.1	0.5	1511.4	−0.9	1511.2	−2.8
$1_1 2_1$	3060.4	−1.5	2208.2	−1.0	2361.8	2.3
$1_1 2_2$	4020.4	0.2	2900.9	3.5	3197.9	−0.7
$1_1 2_3$	4969.4	0.0	3589.6	−0.9	4032.0	0.5
$1_1 2_4$	5910.9	1.3	4274.2	−0.3	4859.3	1.3
$1_1 2_5$	6840.9	0.1	4954.7	−0.4	5680.2	1.9
$1_1 2_6$	7763.4	0.4	$\cdots$	$\cdots$	6491.7	−0.5
$1_1 2_7$	8675.1[b]	−1.0	$\cdots$	$\cdots$	7297.9	−2.0

[a]Energies from Reference [24].
[b]Corrected value from Reference [24].

6.2.8 Renner–Teller and Spin-Orbit Coupling in AsH_2

After R. N. Dixon, G. Duxbury, and H. M. Lamberton, *Proc. Roy. Soc. A*, 305:271, 1968 [16], reproduced with permission from The Royal Society of Chemistry.

Introduction

The valence angles of the ground states of hydrides of elements in the third row of the periodic table were first studied by Dixon, Duxbury, and Lamberton [20] in 1966. In their first study they reported the observation of six absorption bands of AsH_2 and eight bands of AsD_2 between 3900 and 4900 Å. Following on from this, in 1968, Dixon et al. [16] gave a detailed analysis of three of the bands of the AsH_2 spectrum and an analysis of the vibrational structure.

Experimental

Arsine was prepared by the reaction of an alkaline solution of arsenic oxide and potassium borohydride with concentrated sulphuric acid at 0°C in an atmosphere of 100 mm Hg. Perdeuteroarsine was prepared under similar conditions by reacting D_2O with aluminium arsenide, obtained by ignition of an intimate mixture of aluminium and arsenic. The purity of the arsine and perdeuteroarsine were checked by their infrared spectra.

AsH$_2$ and AsD$_2$ radicals were produced by photolysis of arsine and perdeuteroarsine at 20 mm Hg pressure with a 3000 J photolysis flash. The half-life of the radicals was about 100 μs, and all spectra were photographed with 50 μs delay time. The photolysis resulted in the formation of an arsenic mirror on the walls of the reaction vessel, which was cleaned with nitric acid every ten flashes. The initial plates were taken in the second order of a 21 ft. Eagle mounting concave grating spectrograph using two traversals through a 90 cm absorption tube. This was only irradiated over the central 60 cm, thus minimizing deposition of arsenic on the end windows.

The stronger bands of the AsH$_2$ spectrum were rephotographed through the same absorption cell in the 12th and 13th orders of a 3.4 m Jarrell-Ash Ebert spectrograph. The long wavelength end of the AsH$_2$ spectrum was also photographed in the 8th and 11th orders of the Ebert spectrograph using a multiple traversal absorption cell with path lengths of up to 6 m. All spectra were taken with slit widths of 70 or 100 μm on Kodak I-O or I-F plates, with exposures ranging from 40 to 150 flashes. An iron hollow cathode lamp was used for wavelength calibration, and the bands were measured by means of a photoelectric comparator.

Results and Vibrational Analysis

The spectra of AsH$_2$ and AsD$_2$ each consist of a single long progression of bands, with spacings that may be assigned to the upper-state bending frequencies, showing that there is a considerable change in the bending angle in the electronic transition. In the preliminary work, the origin of the band system was predicted to lie at ca. 19905 cm^{-1} by extrapolation of the hydrogen to deuterium isotope shifts. This was confirmed by the observation of the 0–0 band of the AsH$_2$ spectrum at the predicted frequency by using the multiple traversal absorption cell.

The band origins are given in Table 6.11 and are represented to within ± 1 cm^{-1} by the equations:

$$\text{AsH}_2: \quad \nu = 19905.9 + 854.5v'_2 - 3.1v'^2_2 \quad (\text{cm}^{-1}) \tag{6.5}$$
$$\text{AsD}_2: \quad \nu = 19904.9 + 618.0v'_2 - 2.1v'^2_2 \quad (\text{cm}^{-1}).$$

Table 6.11: Vacuum Wave-Numbers of the Band Heads of AsH$_2$ and AsD$_2$

Band	AsH$_2$		AsD$_2$	
	G	ΔG	G	ΔG
$(0,0,0) - (0,0,0)$	19905.5			
		851.4		
$(0,1,0) - (0,0,0)$	20756.9			
		845.4		
$(0,2,0) - (0,0,0)$	21602.3		21132.2	
		839.6		606.6
$(0,3,0) - (0,0,0)$	22441.9		21738.8	
		832.5		603.6
$(0,4,0) - (0,0,0)$	23274.4		22342.0	
		826.5		600.1
$(0,5,0) - (0,0,0)$	24100.9		22942.1	
		820.5		597.1
$(0,6,0) - (0,0,0)$	24921.4		23539.2	
				590.2
$(0,7,0) - (0,0,0)$			24129.4	
				585.8
$(0,8,0) - (0,0,0)$			24715.2	
				580.6
$(0,9,0) - (0,0,0)$			25295.8	

Evaluation of the rotational and spin coupling constants

Errors in the trial rotational constants have been refined by using the deviations between the observed and calculated values of $F_1(N, K_a, K_c) + F_2(N, K_a, K_c)$, and the spin coupling constants from $F_1(N, K_a, K_c) - F_2(N, K_a, K_c)$. In both cases, the coefficients for the least squares analysis were derived from the Hund's case (b) Equations (6.2) and (6.3). This method gave fairly rapid convergence for the ground state, which has the smaller spin-rotation constants, but more iterations were required for the upper state. In both states the four independent fourth power centrifugal distortion constants were chosen to be τ_{aaaa}, τ_{bbbb}, τ_{aabb}, and τ_{abab}.

Constants for the ground state

The assigned lines in the ν_2', $2\nu_2'$, and $3\nu_2'$ bands led to a total of approximately 170 distinct combination differences for the lower state, the majority of which have been determined from all three bands. The molecular constants obtained from the means of these differences are given in Table 6.12, and lead to a standard deviation of 0.11 cm^{-1}. This is comparable to the discrepancies obtained between the combination branches of different bands, and is an indication of the experimental errors.

Table 6.12: Molecular Constants of the (0,0,0) Level of the $\tilde{X}(^2B_1)$ State of AsH$_2$

Parameter	Value	Unit
A_0	7.5486 ± 0.0043	cm^{-1}
B_0	7.1624 ± 0.0045	cm^{-1}
C_0	3.6166 ± 0.0030	cm^{-1}
I_a	2.2337 ± 0.0013	a.m.u.Å^2
I_b	2.3542 ± 0.0014	a.m.u.Å^2
I_c	4.6623 ± 0.0014	a.m.u.Å^2
$\Delta = (I_c - I_a - I_b)$	0.0744	a.m.u.Å^2
τ_{aaaa}	-0.00446 ± 0.00034	cm^{-1}
τ_{bbbb}	-0.00317 ± 0.00045	cm^{-1}
τ_{aabb}	0.00219 ± 0.00086	cm^{-1}
τ_{abab}	-0.00057 ± 0.00042	cm^{-1}
κ	0.8036	
ϵ_{aa}	-1.050 ± 0.023	cm^{-1}
ϵ_{bb}	-0.399 ± 0.025	cm^{-1}
ϵ_{cc}	0.005 ± 0.027	cm^{-1}
$^*r_0(\text{AsH})$	1.518	Å
$^*\theta_0$	$99°44'$	

*Determined from I_a and I_b.

Constants for the excited state

The molecular constants and assignments obtained for the $v_2' = 1$, 2, and 3 levels are compiled in Tables 6.13 and 6.14. For $v_2' = 3$, the number of levels that give good ground state combinations appear to be perturbed by 1 cm^{-1} or more. These levels were omitted from the least squares treatment, and the standard deviation between the observations and calculations was 0.37 cm^{-1} on the remaining levels. The poorer fit to the $v_2' = 3$ level compared with $v_2' = 1$ and 2 indicates either that the model Hamiltonian is not capable of representing the rotational structure, or that there are extensive minor perturbations of the rotational levels.

Table 6.13: Rotational Parameters for Levels of the $\tilde{A}(^2A_1)$ State of AsH_2

Parameter	Level			Unit
	(0,1,0)	(0,2,0)	(0,3,0)	
A_0	19.482 ± 0.014	22.504 ± 0.018	26.421 ± 0.040	cm^{-1}
B_0	4.973 ± 0.007	5.018 ± 0.010	5.116 ± 0.030	cm^{-1}
C_0	3.708 ± 0.004	3.664 ± 0.010	3.693 ± 0.031	cm^{-1}
$\Delta = (I_c - I_a - I_b)$	0.291	0.492	0.632	a.m.u.Å^2
τ_{aaaa}	0.1669 ± 0.0028	-0.3237 ± 0.0053	-0.565 ± 0.012	cm^{-1}
τ_{bbbb}	-0.0013 ± 0.0005	-0.0006 ± 0.0012	-0.006 ± 0.003	cm^{-1}
τ_{aabb}	0.0043 ± 0.0006	0.0062 ± 0.0006	0.007 ± 0.002	cm^{-1}
$*\tau_{abab}$	(0)	(0)	(0)	cm^{-1}
H_k	0.00014 ± 0.00001	0.00051 ± 0.00003	0.00106 ± 0.00005	cm^{-1}
κ	-0.8396	-0.8563	-0.8748	
ϵ_{aa}	6.17 ± 0.05	8.44 ± 0.06	10.91 ± 0.10	cm^{-1}
$*\epsilon_{bb}$	(0)	(0)	(0)	cm^{-1}
ϵ_{cc}	0.15 ± 0.02	-0.13 ± 0.03	-0.23 ± 0.08	cm^{-1}
η_{aaaa}	0.033 ± 0.002	-0.068 ± 0.002	-0.095 ± 0.004	cm^{-1}
$T_{0,v_2,0}$	20759.74 ± 0.07	21605.43 ± 0.07	22444.94 ± 0.07	cm^{-1}

*Constrained to zero.

Rotational assignments

Table 6.14: Vacuum Wave-Numbers And and Rotational Assignments for the $(0, v_2, 0)$–$(0,0,0)$ Bands of AsH_2, with $v_2 = 1$, 2, and 3

Branch	Rotational Transition	Calculated Relative Intensity[†]	$v_2 = 1$	$v_2 = 2$	$v_2 = 3$
$^R R_{0,N}$	$1_{00} - 0_{00}$	3.00	20785.22	21633.92	22477.19
			81.04[s]	28.53[s]	70.22[s]
	$2_{11} - 1_{01}$	1.43	–	21641.56	22485.01
			–	39.50	82.79
	$3_{12} - 2_{02}$	3.36	20802.16	21651.05	22494.75
			0.57	49.10	93.09
	$4_{13} - 3_{03}$	0.77	–	–	–
	$5_{14} - 4_{04}$	1.67	20827.07	–	22519.91
			–	–	19.02
	$6_{15} - 5_{05}$	0.41	–	–	–
	$7_{16} - 6_{06}$	0.93	20858.15	–	–
			57.26	–	–
$^R R_{1,N}$	$2_{21} - 1_{11}$	4.26	20837.84	21695.43	22549.95[p]
			28.00	83.05	34.51
	$3_{22} - 2_{12}$	1.45	20844.93	21702.43	22557.21
			39.12	696.79*	51.67*
	$4_{23} - 3_{13}$	3.77	20853.76	21711.24	22566.20
			49.49	06.52	61.85
	$5_{24} - 4_{14}$	1.00	20863.72	21721.53*	–
			–	17.98	–
	$6_{25} - 5_{15}$	2.29	20875.84	21733.30*	–
			72.62	29.76	–
$^R R_{1,N-1}$	$2_{20} - 1_{10}$	1.45	20834.29	21692.01	22546.51[p]
			24.36	79.50*	30.87
	$3_{21} - 2_{11}$	4.65	20834.87	21692.28	22546.94
			28.80	85.61	40.93[p]
	$4_{22} - 3_{12}$	1.09	–	–	22548.78
			–	–	43.64
	$5_{23} - 4_{13}$	1.68	20840.89	21598.31	22553.04
			–	93.26	48.30
$^R R_{2,N-1}$	$3_{31} - 2_{21}$	2.07	20909.78	21780.43	22651.24
			–	61.09	27.98
	$4_{32} - 3_{22}$	5.69	20910.97	21781.47	22652.37*
			00.44	69.73	40.94
	$5_{33} - 4_{23}$	1.47	20913.87	21784.55	22655.37*
			05.48	74.54	46.06*

Table 6.14: Vacuum Wave-Numbers And and Rotational Assignments for the $(0, v_2, 0)-(0,0,0)$ Bands of AsH$_2$, with $v_2 = 1$, 2, and 3

Branch	Rotational Transition	Calculated Relative Intensity[†]	$v_2 = 1$	$v_2 = 2$	$v_2 = 3$
	$6_{34} - 5_{24}$	3.01	–	21789.00	22660.18
			–	80.95	52.37*
$^R R_{2,N-2}$	$3_{30} - 2_{20}$	5.18	20906.61	21777.24	22648.01
			890.86	57.97	24.84
	$4_{31} - 3_{21}$	1.32	–	21771.48	22642.34
			–	59.37	30.84
	$5_{32} - 4_{22}$	2.16	20896.00	21766.30*	22637.24
			86.91	56.43	27.64
$^R R_{3,N-2}$	$4_{41} - 3_{31}$	6.71	21001.50	21888.10	22777.91
			980.27	62.57	47.60
	$5_{42} - 4_{32}$	1.79	20997.04	21833.31	22773.07
			81.52*	66.00*	55.12
	$6_{43} - 5_{33}$	3.58	20993.94	21880.20	22770.02
			81.52*	66.11*	55.22
$^R R_{3,N-3}$	$4_{40} - 3_{30}$	1.90	20998.72	21885.33	22775.06
			77.70	60.00	45.05
	$5_{41} - 4_{31}$	3.38	20987.61	21873.91	22763.64
			72.44	56.91	45.92
	$6_{42} - 5_{32}$	0.49	–	21862.57*	–
			–	48.14*	–
$^R R_{4,N-3}$	$5_{51} - 4_{41}$	2.09	21110.24	22014.37	22914.05
			084.02*	21983.39	888.32*
	$6_{52} - 5_{42}$	4.41	21100.35	22004.14	22914.05
			080.26	21981.46*	889.71
	$7_{53} - 6_{43}$	0.84	–	–	22901.97*
			–	–	884.96*
	$8_{54} - 7_{44}$	1.13	21084.02*	–	–
			–	–	–
$^R Q_{4,N-3}$	$5_{51} - 5_{42}$	$(F_1\text{-}F_2)$	–	–	22860.32*
$^R R_{4,N-4}$	$5_{50} - 4_{40}$	5.64	21107.96	22012.03	22922.11
			082.09	21981.46	886.34
	$6_{51} - 5_{41}$	0.96	–	–	22905.45
			21073.11	–	881.71
	$7_{52} - 6_{42}$	1.08	–	–	22888.32*
			–	–	68.01
$^R S_{4,N-4}$	$6_{51} - 4_{40}$	$(F_2\text{-}F_1)$	–	–	22951.58^s
$^R R_{5,N-4}$	$6_{61} - 5_{50}$	5.22	21233.64	22155.99	23086.27
			03.03*	20.36	45.19

Table 6.14: Vacuum Wave-Numbers And and Rotational Assignments for the $(0, v_2, 0)$–$(0,0,0)$ Bands of AsH_2, with $v_2 = 1$, 2, and 3

Branch	Rotational Transition	Calculated Relative Intensity[†]	$v_2 = 1$	$v_2 = 2$	$v_2 = 3$
	$7_{62} - 6_{52}$	1.10	−	22140.48	23070.64
			−	13.17*	40.61
	$8_{63} - 7_{53}$	1.63	21203.04*	22125.39	−
			−	02.85	−
$^R S_{5,N-4}$	$7_{62} - 5_{51}$	$(F_2\text{-}F_1)$	−	−	23119.37^s
$^R R_{5,N-5}$	$6_{60} - 5_{50}$	1.65	21231.83	22154.18	23084.42
			01.58	18.80*	43.71
	$7_{61} - 6_{51}$	2.39	21211.18	22133.17	23063.09
			187.38	06.34	33.77
$^P P_{1,N-1}$	$0_{00} - 1_{10}$	0.93	−	21590.77*	−
			−	89.96*	−
	$1_{01} - 2_{11}$	3.73	20729.03	21574.74	22414.49
			27.99	73.59	13.31*
	$2_{02} - 3_{12}$	1.17	−	21559.07*	−
			−	57.42*	−
	$3_{03} - 4_{13}$	2.68	20698.80	21544.60*	22384.89
			97.02	43.09	82.61
	$4_{04} - 5_{14}$	0.68	−	21531.74*	−
			−	29.22*	−
	$5_{05} - 6_{15}$	1.57	−	21518.60*	−
			−	16.96*	−
$^P P_{2,N-1}$	$1_{11} - 2_{21}$	1.23	−	21592.25	22435.63
			−	85.73	26.92
	$2_{12} - 3_{22}$	3.51	20727.08	21575.86	22419.55
			24.07	72.49	16.67
	$3_{13} - 4_{23}$	1.02	−	21559.07	−
			−	56.72	−
	$4_{14} - 5_{24}$	2.46	20695.61	21544.14	22387.70
			93.48	41.85	85.71
$^P P_{2,N-2}$	$1_{10} - 2_{20}$	4.80	20741.84	21590.56	22433.91
			35.74	83.34	25.04
	$2_{11} - 3_{21}$	1.79	−	21565.69	22413.32*
			−	65.84	09.48
	$3_{12} - 4_{22}$	4.66	20700.60	21549.52	22393.27
			697.02	45.79	89.79
	$4_{13} - 5_{23}$	1.00	−	21532.88	−
			−	29.22	−
	$5_{14} - 6_{24}$	1.82	−	21516.96	22359.93

Table 6.14: Vacuum Wave-Numbers And and Rotational Assignments for the $(0, v_2, 0)$–$(0,0,0)$ Bands of AsH_2, with $v_2 = 1$, 2, and 3

Branch	Rotational Transition	Calculated Relative Intensity†	$v_2 = 1$	$v_2 = 2$	$v_2 = 3$
			—	14.48	56.69
$^PP_{3,N-2}$	$2_{21} - 3_{31}$	5.24	20764.42	21622.05*	22476.53[p]
			52.39	07.40*	58.99
	$3_{22} - 4_{32}$	1.58	—	21600.47	22455.36
			20735.29	592.25	47.79[p]
	$4_{23} - 5_{33}$	3.86	20722.82	21580.32	22435.28
			16.39	73.59	28.82
	$5_{24} - 6_{34}$	0.94	—	21561.38*	—
			—	55.41*	—
	$6_{25} - 7_{35}$	1.92	—	21544.14	—
			—	38.34	—
$^PP_{3,N-3}$	$2_{20} - 3_{30}$	1.72	20761.60	21619.08*	22473.63[p]
			49.91	04.89*	56.48
	$3_{21} - 4_{31}$	4.46	20734.02	21591.43	22446.11*
			26.63	83.34	38.81[p]
	$4_{22} - 5_{32}$	1.12	—	21563.59	22417.57
			—	56.72	11.76
	$5_{23} - 6_{33}$	1.88	20681.03	21538.34	22393.13
			—	31.74	86.66
$^PP_{4,N-3}$	$3_{31} - 4_{41}$	1.78	20806.62	21677.36	22548.06
			788.86	55.96	22.77
	$4_{32} - 5_{42}$	4.35	20779.67	21650.16	22521.08
			67.42	36.57	07.73
	$5_{33} - 6_{43}$	1.05	20754.07	21624.28*	22495.25
			—	—	83.93*
	$6_{34} - 7_{44}$	2.03	20729.48	21600.00	22471.16
			—	589.96	61.45*
$^PP_{4,N-4}$	$3_{30} - 4_{40}$	5.05	20804.30	21674.93	22545.73
			786.81	53.98	20.80
	$4_{31} - 5_{41}$	1.14	20771.18	21641.56	22512.57
			—	28.53*	499.72
	$5_{32} - 6_{42}$	1.96	20736.95	21607.40*	22478.24
			26.63	596.25	67.43
$^PP_{5,N-4}$	$4_{41} - 5_{51}$	4.63	20868.61	21755.25	22645.18
			45.37	27.75	13.01
	$5_{42} - 6_{52}$	1.12	—	21722.80*	22612.40
			20819.92	03.68	592.88
	$6_{43} - 7_{53}$	2.13	20804.81	21691.20	22581.10

Table 6.14: Vacuum Wave-Numbers And and Rotational Assignments for the $(0, v_2, 0)$–$(0,0,0)$ Bands of AsH_2, with $v_2 = 1, 2,$ and 3

Branch	Rotational Transition	Calculated Relative Intensity[†]	$v_2 = 1$	$v_2 = 2$	$v_2 = 3$
			791.36	75.21	64.72
$^PP_{5,N-5}$	$4_{40} - 5_{50}$	1.48	20866.74	21753.47	22643.39
			43.85	26.20*	11.52
	$5_{41} - 6_{51}$	2.64	20828.80	21715.22	22604.90
			13.16	696.79	586.04
$^PP_{6,N-5}$	$5_{51} - 6_{61}$	1.19	20948.05	21852.10	22762.06
			19.85	19.25	24.32*
	$6_{52} - 7_{62}$	2.33	20910.57	21814.47	22724.32*
			–	790.17	698.33
	$7_{53} - 8_{63}$	0.43	–	–	22687.54
$^PP_{6,N-6}$	$5_{50} - 6_{60}$	3.50	20946.65	21850.85	22760.75
			18.78	18.22	23.02
	$6_{51} - 7_{61}$	0.65	20904.40	–	22717.96
			–	–	692.46
$^PP_{7,N-6}$	$6_{61} - 7_{71}$	2.49	21042.14	21964.42	22894.63
			09.56	26.89	51.63
	$7_{62} - 8_{72}$	0.49	–	–	22851.95
			–	–	20.12*
$^PQ_{7,N-6}$	$7_{62} - 7_{71}$	$(F_2\text{-}F_1)$	–	–	22927.83^s
$^PP_{7,N-7}$	$6_{60} - 7_{70}$	0.82	21041.14	21963.55	22893.62
			–	26.17	50.91
	$7_{61} - 8_{71}$	1.34	20994.74	21916.88	22846.69
			–	888.32	15.75
$^RQ_{0,N}$	$1_{11} - 1_{01}$	1.43	22773.39	22622.05*	22465.67
			–	26.17	50.91
	$2_{12} - 2_{02}$	8.61	22770.68	21619.41	22463.18
			68.37*	17.24*	61.45
	$3_{13} - 3_{03}$	3.85	20769.14*	–	–
			67.42*	–	–
	$4_{14} - 4_{04}$	12.98	20768.37*	21616.87*	22460.53*
			67.42*	15.86*	59.73*
	$5_{15} - 5_{05}$	4.40	20768.37*	(head)	(head)
			67.42*	(21615.8*)	(22460.1*)
	$6_{16} - 6_{06}$	12.48	20768.77*	–	–
			68.37*	–	–
	$7_{17} - 7_{07}$	3.71	20769.68*	–	–
			69.68*	–	–

Table 6.14: Vacuum Wave-Numbers And and Rotational Assignments for the $(0, v_2, 0)$–$(0,0,0)$ Bands of AsH_2, with $v_2 = 1$, 2, and 3

Branch	Rotational Transition	Calculated Relative Intensity[†]	$v_2 = 1$	$v_2 = 2$	$v_2 = 3$
	$8_{18} - 8_{08}$	9.43	20770.68*	–	–
			71.18*	–	–
$^R Q_{1,N}$	$2_{20} - 2_{12}$	0.63	–	–	–
	$3_{21} - 3_{13}$	3.42	22819.12	21677.35	22531.99
			14.35	71.15	26.49^p
	$4_{22} - 4_{14}$	1.38	22822.19	21679.50	22534.34
			–	74.39	30.06
	$5_{23} - 5_{15}$	4.05	22826.62	21683.92	22538.75
			23.14	80.14	35.17
	$6_{24} - 6_{16}$	1.14	20833.33	–	–
	$7_{25} - 7_{17}$	2.63	20842.43	–	–
			39.84	–	–
$^R Q_{1,N-1}$	$2_{21} - 2_{11}$	2.07	20809.36	21666.97	21521.48^p
			799.25	54.29	05.81
	$3_{22} - 3_{12}$	1.57	20801.50	21658.94	22513.73
			795.00	51.82	07.21^p
	$4_{23} - 4_{13}$	6.35	20795.60	21653.11	22508.08
			90.20	47.28	02.55
	$5_{24} - 5_{14}$	2.21	20791.36	21648.78	–
			86.81	43.76	–
	$6_{25} - 6_{15}$	6.13	20788.44	21646.01	–
			83.99	41.26	–
	$7_{26} - 7_{16}$	1.75	–	–	–
			20882.59*	–	–
	$8_{27} - 8_{17}$	4.22	20787.64	21643.76	–
			82.59*	39.50	–
$^R Q_{2,N-1}$	$3_{30} - 3_{22}$	1.75	20877.44	21748.09	22618.86
			61.90	28.90	595.85
	$4_{31} - 4_{23}$	1.13	–	21741.67	22612.40
				29.76	00.88
	$5_{32} - 5_{24}$	4.37	20866.74	21737.14	22608.16
			58.15	27.75	598.85*
	$6_{33} - 6_{25}$	1.50	20864.24	–	22606.45
			58.15	–	598.85*
	$7_{34} - 7_{26}$	4.03	20863.72	–	–
			57.00	–	–
	$8_{35} - 8_{27}$	1.08	20864.24	–	–
			59.25	–	–

Table 6.14: Vacuum Wave-Numbers And and Rotational Assignments for the $(0, v_2, 0)$–$(0,0,0)$ Bands of AsH_2, with $v_2 = 1$, 2, and 3

Branch	Rotational Transition	Calculated Relative Intensity[†]	$v_2 = 1$	$v_2 = 2$	$v_2 = 3$
$^RQ_{2,N-2}$	$3_{31} - 3_{21}$	0.40	20867.43	–	22608.79
			51.94	–	–
	$4_{32} - 4_{22}$	2.48	20853.06	21723.54	22594.43
			41.92	11.24	82.40
	$5_{33} - 5_{23}$	1.05	–	–	22582.59
			–	–	72.56
	$6_{34} - 6_{24}$	2.90	–	21701.68	22572.82
			–	692.59	63.96
$^RQ_{3,N-2}$	$4_{40} - 4_{32}$	0.42	–	–	22731.25
			–	–	01.54
	$5_{41} - 5_{33}$	2.32	20942.55	21828.88	22718.66
			–	11.53	00.72
	$6_{42} - 6_{34}$	0.94	20932.21	–	22708.26
			19.85	–	693.52
	$7_{43} - 7_{35}$	2.72	20923.60	21809.92	22699.87
			12.31	797.41	87.24
$^RQ_{3,N-3}$	$4_{41} - 4_{31}$	0.79	–	–	22721.98
			–	–	691.95
	$5_{42} - 5_{32}$	0.42	–	–	22701.06
			–	–	682.81
$^RQ_{4,N-3}$	$5_{50} - 5_{42}$	0.83	–	–	22864.01
			–	–	28.29
	$6_{51} - 6_{43}$	0.46	–	–	22845.31
			–	–	20.98
	$7_{52} - 7_{44}$	1.54	–	–	22829.00
			–	–	08.69
$^RQ_{4,N-4}$	$5_{51} - 5_{41}$	0.18	–	–	–
			–	–	–
	$6_{52} - 6_{42}$	0.72	–	–	22828.48
			–	–	04.09
$^PQ_{1,N}$	$1_{01} - 1_{11}$	4.26	20757.57*	21603.28*	22442.92*
			56.55*	02.40*	41.94*
	$2_{02} - 2_{12}$	2.26	–	–	22442.05*
			–	–	41.65*
	$3_{03} - 3_{13}$	8.67	20756.91	21602.60	22443.00*
			56.55*	02.32*	41.94*
	$4_{04} - 4_{14}$	3.30	20758.13	21603.99*	–
			57.73*	03.46*	–

Table 6.14: Vacuum Wave-Numbers And and Rotational Assignments for the $(0, v_2, 0)$–$(0,0,0)$ Bands of AsH_2, with $v_2 = 1$, 2, and 3

Branch	Rotational Transition	Calculated Relative Intensity[†]	$v_2 = 1$	$v_2 = 2$	$v_2 = 3$
	$5_{05} - 5_{15}$	10.52	20760.09	21605.93*	22446.19*
			59.78	05.62	–
	$6_{06} - 6_{16}$	3.50	20762.45	21608.40	–
			62.15	08.17*	–
	$7_{07} - 7_{17}$	9.83	20765.10	21611.19	22451.18*
			64.73	10.86	–
	$8_{08} - 8_{18}$	2.90	20767.77	21614.82	–
			66.47	14.07	–
	$9_{09} - 9_{19}$	7.21	20770.68*	–	–
			68.37*	–	–
$^{P}Q_{2,N-1}$	$2_{11} - 2_{21}$	0.68	–	–	22455.36*
			–	–	51.70
	$3_{12} - 3_{22}$	3.17	20758.59	20607.40*	22451.18
			55.54	04.20*	48.28
	$4_{13} - 4_{23}$	1.20	–	–	–
			–	–	–
	$5_{14} - 5_{24}$	3.55	20754.51	–	22445.02
			52.39	–	44.76
	$6_{15} - 6_{25}$	1.08	–	–	–
			–	–	–
	$7_{16} - 7_{26}$	2.84	20756.55	–	–
			54.07	–	–
$^{T}R_{0,N}$	$3_{30} - 2_{02}$	1.02	21921.16	21791.68	22662.48
			06.61	73.70*	40.70
	$4_{31} - 3_{03}$	0.58	–	–	22770.60
			–	–	60.18*
	$5_{32} - 4_{04}$	1.84	20939.61	20809.92	22680.94
			32.21	01.70	72.88
	$6_{33} - 5_{05}$	0.54	–	–	22693.52*
			–	–	87.24*
$^{T}R_{1,N-1}$	$4_{40} - 3_{12}$	0.43	–	–	–
			–	–	–
	$5_{41} - 4_{13}$	2.56	21015.37	21901.70	22791.51
			00.95	855.33	74.46
	$6_{42} - 5_{14}$	0.89	–	–	22795.61
			–	–	81.78
	$7_{43} - 6_{15}$	2.19	21025.53	21912.01	22801.71
			15.37	00.10	90.16

Table 6.14: Vacuum Wave-Numbers And and Rotational Assignments for the $(0, v_2, 0)$–$(0,0,0)$ Bands of AsH_2, with $v_2 = 1$, 2, and 3

Branch	Rotational Transition	Calculated Relative Intensity[†]	$v_2 = 1$	$v_2 = 2$	$v_2 = 3$
$^TR_{2,N-1}$	$5_{51} - 4_{23}$	0.16	–	–	–
			–	–	–
	$6_{52} - 5_{24}$	1.46	–	22048.23	22958.23
			–	27.33	35.65
$^TR_{2,N-2}$	$5_{50} - 4_{22}$	1.02	21123.10	22027.33	22937.33
			098.56	21998.02	02.86
	$6_{51} - 5_{23}$	0.78	21119.02	–	22932.65
			00.78	–	09.28
	$7_{52} - 6_{24}$	2.39	21117.13	–	22930.67
			02.05	–	11.33
$^TS_{2,N-2}$	$6_{51} - 5_{23}$	(F_2-F_1)	–	–	22966.83[s]
$^TR_{3,N-2}$	$6_{61} - 5_{33}$	0.49	–	–	23119.37
			–	–	079.25
$^TR_{3,N-3}$	$6_{60} - 5_{32}$	0.22	–	–	–
	$7_{61} - 6_{33}$	1.66	–	22151.71	23089.59
			–	33.17*	60.59
$^PR_{1,N-1}$	$2_{02} - 1_{10}$	0.41	–	–	–
	$3_{03} - 2_{11}$	1.98	–	–	23458.15
			–	–	56.48
$^PR_{2,N-2}$	$3_{12} - 2_{20}$	2.06	–	–	22480.23
			–	–	77.19*
	$4_{13} - 3_{21}$	1.20	–	–	–
	$5_{14} - 4_{22}$	3.68	–	–	22476.53
$^PR_{3,N-3}$	$4_{22} - 3_{30}$	0.58	–	–	–
	$5_{23} - 4_{31}$	3.31	–	21670.48	22525.26
			–	64.85	19.91
$^PR_{4,N-3}$	$5_{33} - 4_{41}$	0.25	–	–	–
	$6_{34} - 5_{42}$	1.42	–	21744.91	22616.00
			–	35.27	06.45
$^PR_{4,N-4}$	$5_{32} - 4_{40}$	1.16	–	21751.17	22622.18
			–	39.91	11.15
	$6_{33} - 5_{41}$	0.73	–	–	22608.16
			–	–	599.18
$^PR_{5,N-5}$	$6_{42} - 5_{50}$	0.26	–	–	–
	$7_{43} - 6_{51}$	1.41	–	–	22716.94
			–	–	03.43
$^TQ_{0,N}$	$3_{31} - 3_{03}$	0.18	–	–	–
	$4_{32} - 4_{04}$	1.20	20896.55	21767.05	22637.99

Table 6.14: Vacuum Wave-Numbers And and Rotational Assignments for the $(0, v_2, 0)$–$(0,0,0)$ Bands of AsH_2, with $v_2 = 1$, 2, and 3

Branch	Rotational Transition	Calculated Relative Intensity†	$v_2 = 1$	$v_2 = 2$	$v_2 = 3$
			86.87	56.43	27.64
	$5_{33} - 5_{05}$	0.56	–	–	–
			–	–	–
	$6_{34} - 6_{06}$	1.90	20904.33	21774.91	22646.19
			898.32	68.02	39.42
	$7_{35} - 7_{07}$	0.61	20911.10	–	–
			05.48	–	–
	$8_{36} - 8_{08}$	1.62	20919.23	–	–
			14.50	–	–
$^T Q_{1,N-1}$	$4_{41} - 4_{13}$	0.60	–	–	–
	$5_{42} - 5_{14}$	0.48	–	–	–
	$6_{43} - 6_{15}$	2.04	20964.94*	21851.36	22741.21
			53.09	37.95	27.37
	$7_{44} - 7_{16}$	0.75	–	–	22740.11
			–	–	28.45
	$8_{45} - 8_{17}$	2.12	20964.94*	–	–
			55.51	–	–
$^T Q_{2,N-2}$	$5_{51} - 5_{23}$	0.17	–	–	–
	$6_{52} - 6_{24}$	1.09	21057.22	–	22870.85
			37.81	–	47.28

Branches are designated by $^{\Delta K_a} \Delta N_{K_a'', K_c''}$.
Entries are $F_1 - F_1$ (upper) and $F_2 - F_2$ (lower).
†Calculations for Hund's case (b).
*Denotes blended line.
^{s}Satellite transition $F_2 - F_1$.
^{p}Perturbed upper level.

6.3 Bibliography

[1] K. Saito and K. Obi. The first observation of the $\tilde{A}^1 B_1 - \tilde{X}^1 A_1$ transition of GeH_2 and GeD_2 radicals. *Chemical Physics Letters*, 215(1):193–198, 1993.

[2] K. Obi, M. Fukushima, and K. Saito. Detection of silylene and germylene radicals by laser-induced fluorescence. *Applied Surface Science*, 79:465–470, 1994.

[3] J. Karolczak, W. W. Harper, R. S. Grev, and D. J. Clouthier. The structure, spectroscopy, and excited state predissociation dynamics of GeH_2. *The Journal of Chemical Physics*, 103(8):2839–2849, 1995.

[4] A. Campargue and R. Escribano. Room temperature absorption spectroscopy of GeH_2 near 585 nm. *Chemical Physics Letters*, 315(5–6):397–404, 1999.

[5] M. Fukushima, S. Mayama, and K. Obi. Jet spectroscopy and excited state dynamics of SiH_2 and SiD_2. *The Journal of Chemical Physics*, 96(1):44–52, 1992.

[6] K. Saito and K. Obi. The excited state dynamics of the $\tilde{A}^1 B_1$ state of GeH_2 and GeD_2 radicals. *Chemical Physics*, 187(3):381–389, 1994.

[7] T. C. Smith, D. J. Clouthier, W. Sha, and A. G. Adam. Laser optogalvanic and jet spectroscopy of germylene (GeH_2): New spectroscopic data for an important semiconductor growth intermediate. *The Journal of Chemical Physics*, 113(21):9567–9576, 2000.

[8] D. R. Lyons, A. L. Schawlow, and G.-Y. Yan. Doppler-free radiofrequency optogalvanic spectroscopy. *Optics Communications*, 38(1):35–38, 1981.

[9] R. Vasudev and R. N. Zare. Laser optogalvanic study of HCO $\tilde{A}$ state predissociation. *The Journal of Chemical Physics*, 76(11):5267–5270, 1982.

[10] R. D. May and P. H. May. Solid-state radio frequency oscillator for optogalvanic spectroscopy: Detection of nitric oxide using the 2-0 overtone transition. *Review of Scientific Instruments*, 57(9):2242–2245, 1986.

[11] W. W. Harper and D. J. Clouthier. Reinvestigation of the HSiCl electronic spectrum: Experimental reevaluation of the geometry, rotational constants, and vibrational frequencies. *The Journal of Chemical Physics*, 106(23):9461–9473, 1997.

[12] P. R. Bunker, R. A. Phillips, and R. J. Buenker. An ab initio study of the rotation-vibration energy levels of GeH_2 in the $\tilde{X}^1 A_1$ state. *Chemical Physics Letters*, 110(4):351–355, 1984.

[13] G. R. Smith and W. A. Guillory. Products of the vacuum-ultraviolet photolysis of germane isolated in an argon matrix. *The Journal of Chemical Physics*, 56(4):1423–1430, 1972.

[14] J. C. Barthelat, B. S. Roch, G. Trinquier, and J. Satge. Structure and singlet-triplet separation in simple germylenes GeH_2, GeF_2, and $Ge(CH_3)_2$. *Journal of the American Chemical Society*, 102(12):4080–4085, 1980.

[15] R. N. Dixon and G. Duxbury. Doublet splittings in a non-rigid molecule. *Chemical Physics Letters*, 1(8):330–332, 1967.

[16] R. N. Dixon, G. Duxbury, and H. M. Lamberton. The analysis of a $^2A_1 - {}^2B_1$ electronic band system of the AsH_2 and AsD_2 radicals. *Proceedings of the Royal Society of London. Series A. Mathematical and Physical Sciences*, 305(1481):271–290, 1968.

[17] H. Fujiwara, K. Kobayashi, H. Ozeki, and S. Saito. Submillimeter-wave spectrum of the AsH_2 radical in the $^2B_1$ ground electronic state. *The Journal of Chemical Physics*, 109(13):5351–5355, 1998.

[18] H. Fujiwara and S. Saito. Microwave spectrum of the $AsD_2 \tilde{X}^2 B_1$ radical: Harmonic force field and molecular structure. *Journal of Molecular Spectroscopy*, 192(2):399–405, 1998.

[19] R. A. Hughes, J. M. Brown, and K. M. Evenson. Rotational spectrum of the AsH$_2$ radical in its ground state, studied by far-infrared laser magnetic resonance. *Journal of Molecular Spectroscopy*, 200(2):210–228, 2000.

[20] R. N. Dixon, G. Duxbury, and H. M. Lamberton. Arsenic hydride radicals. *Chemical Communications (London)*, pages 460b–461, 1966.

[21] G. Duxbury and A. Alijah. Effects of spin-orbit coupling on the spin-rotation interaction in the AsH$_2$ radical. In *69th International Symposium on Molecular Spectroscopy*, volume 1, 2014.

[22] H. Lefebvre-Brion and R. W. Field. *The Spectra and Dynamics of Diatomic Molecules*. Academic Press, San Diego, 2004.

[23] D.-F. Zhao. *Cavity ringdown spectroscopy of several radicals*. PhD thesis, University of Science and Technology of China, Hefei, 2009.

[24] S.-G. He and D. J. Clouthier. Laser spectroscopy and dynamics of the jet-cooled AsH$_2$ free radical. *The Journal of Chemical Physics*, 126(15):154312, 2007.

[25] D. Zhao, C. Qin, M. Ji, Q. Zhang, and Y. Chen. Absorption spectra of AsH$_2$ radical in 435-510 nm by cavity ringdown spectroscopy. *Journal of Molecular Spectroscopy*, 256(2):192–197, 2009.

[26] R. A. Grimminger and D. J. Clouthier. Toward an improved understanding of the AsH$_2$ free radical: Laser spectroscopy, ab initio calculations, and normal coordinate analysis. *The Journal of Chemical Physics*, 137(22):224307, 2012.

CHAPTER 7

Astrophysics: The Electronic Spectrum of H_2O^+ and Its Relationships to the Observations Made from Herschel, an ESA Space Observatory

CONTENTS

7.1   INTRODUCTION: THE DISCOVERY OF H_2O^+ BY LEW AND HEIBER

Prior to its discovery in 1973 by Lew and Heiber [1], the electronic spectrum of H_2O^+ had been sought for many years, both at the National Research Council of Canada and elsewhere. Astrophysically, H_2O^+ was expected to be a constituent in comet tails since, according to Whipple [2,3], the comet nucleus was most probably composed largely of ices, including H_2O and H_2O^+.

In addition to the original communication by Lew and Heiber [1], two short letters appeared, by Herzberg and Lew [4] and by Wehinger et al. [5], in which the wavelengths of some H_2O^+ lines were given. Both letters were concerned with the identification of H_2O^+ in the tail of Comet Kohoutek.

Comet Kohoutek was the source of much observational work at NRC in 1973 at radio observatories in Algonquin Park and at Penticton, British Columbia. One of the key discoveries was made in 1973 by Gerhard Herzberg, a Nobel laureate, and Hin Lew. Their identification of the molecular ion H_2O^+ as a constituent of the comet tail was carried out as part of the programme of spectroscopy of the Division of Physics. The unequivocal identification of H_2O^+ in the tail of Comet Kohoutek was the first clear identification that water was a constituent of the comet's nucleus. This major discovery was reported by G. Herzberg and H. Lew [4] in a journal article titled "Tentative identification of the H_2O^+ ion in Comet Kohoutek." This was followed by P. A. Wehinger, S. Wykoff, G. H. Herbig, G. Herzberg, and H. Lew [5].

7.2 OBSERVATIONAL COMETARY SPECTRUM AND LABORATORY SPECTRUM

7.2.1 Spectroscopic Investigations of Fragment Species in the Coma

Paul D. Feldman, Johns Hopkins University, Anita L. Cochran, University of Texas, and Michael R. Combi, University of Michigan [6].[1]

Introduction

The content of the gaseous coma of a comet is dominated by fragment species produced by the photolysis of parent molecules issuing directly from the icy nucleus of the comet. Spectroscopy of these species provides complementary information of the physical state of the coma to that obtained from the observations of the parent species. Extraction of physical parameters requires detailed molecular and atomic data, together with reliable high-resolution spectra and absolute fluxes of the primary source of excitation, the sun. The large database of observations, dating back more than a century, provides a means to assess the chemical and evolutionary diversity of comets.

In 1964, P. Swings delivered the George Darwin lecture [7] on the one–hundredth anniversary of the first reported spectroscopic observation of a comet. In his lecture, he described a century of mainly photographic observations of what were recognised to be fragment species caused by chemical processes acting on the volatile species released from the cometary nuclei in response to solar heating. The features identified in the spectra were the bands of the radicals OH, NH, CN, C_3, C_2, and NH_2; and the ions OH^+, CH^+, CO_2^+, and N_2^+; and the forbidden red doublet of O_I [8,9].

By the time that a book on Comets, by Wilkening and Matthews [10], had been published 18 years later, photoelectric spectroscopy had become quantitative photoelectric spectrometry (A'Hearn [11]), but the inventory of species grew slowly, starting with the sun-grazing Comet Ikeya-Seki (C/1965 S1), and H_2O^+ was identified in Comet Kohoutek (C/1973 E1). In the following two decades, a great deal of information was gathered on the behaviour of fragment species in the Coma.

Molecular ions

Photolytic processes and chemical reactions will ionize molecules in the comae of comets and, as a result, various molecular ions have been detected in the spectra of comets. In the UV and optical regions, these include CH^+, CO_2^+, H_2O^+, N_2^+, and OH^+. Ions have now been detected in the radio spectrum of Comet C/1995 01 (Hale-Bopp), including HCO^+ [12], H_3O^+, and CO^+ [13].

---

[1]From "Spectroscopic Investigations of Fragment Species in the Coma" by P. D. Feldman, A. L. Cochran, and M. R. Combi, in *Comets II*, edited by Michel C. Festou, H. Uwe Keller, and Harold A. Weaver [6]. © 2004 The Arizona Board of Regents. Reprinted by permission of the University of Arizona Press.

The ions show a very different distribution in the coma than do neutrals, since the ions are accelerated tailward by the solar wind. From this behaviour the ionic species are often called "tail" species. However, many of the ionic species are observed relatively close to the nucleus of a comet; for example, the long-slit spectra of CO^+ and CO_2^+ in Comet 1P/Halley given by Umbach et al. [14].

The predominant processes for the production of ions are [15] photodissociation (e.g., $H_2O + h\nu \rightarrow OH^+ + H + e$) and photoionization (e.g., $H_2O + h\nu \rightarrow H_2O^+ + e$). Within the collisional zone (the inner few thousand kilometres of the coma for moderately bright comets), ions can be produced by charge exchange with solar wind protons, electron impact ionization, charge transfer reactions, and proton transfer reactions.

An ion that appears prominently in the red part of the spectrum is H_2O^+. Although it has been observed in cometary spectra for a long time, it was only identified for the first time in the spectra of Comet C/1973 E1 (Kohoutek) [4]). The electronic transition is $\tilde{A}\,^2A_1 - \tilde{X}\,^2B_1$ and is observed from 4000 to 7500 Å. A small part of this range is seen in Figure 7.1. Wegman et al. [16] have run magnetohydrodynamic and chemical simulations of cometary comae and have concluded that for small comets; up to 11% of water molecules are ultimately ionised. H_2O^+ occurs in the spectral bandpass that is most easily accessible to charge coupled device (CCD) detectors, so that there are many observations of H_2O^+ in cometary comae. Lutz et al. [17] have calculated the fluorescence efficiency factors for six of the bands. H_2O^+ is isoelectronic to NH_2, so Arpigny's [18] comment concerning increasing the efficiency factors of NH_2 by a factor of 2 also applies to H_2O^+. Indeed, although the standard reference on the H_2O^+ band [19] used the linear notation, the transitions are more correctly specified in their bent notation (Cochran and Cochran [20]). They have been studied in a high spectral resolution atlas of Comet 122P/de Vico [20].

Bonev and Jockers [21] mapped the distribution of H_2O^+ in Comet C/1989 X1 Austin. They found a strong asymmetry with a relatively flat distribution tailward and a factor of 4 dropoff in the first 10^4 km sunward. The maximum H_2O^+ column density was frequently observed to be shifted tailward.

Outlook

The study of gaseous content of cometary comae has seen much progress during the past two decades. This is due to both enhancements in technology, enabling many more species to be observed, and to a better understanding of the physical processes producing the observed emissions. Spectroscopy continues to be a powerful tool that remains ahead of the laboratory data needed to identify the still large numbers of unexplained features seen in high-resolution spectra, both in the visible [20] and the far-UV [22]. The large database of observations, dating back to more than a century, provides an important tool to assess the chemical and evolutionary diversity of comets.

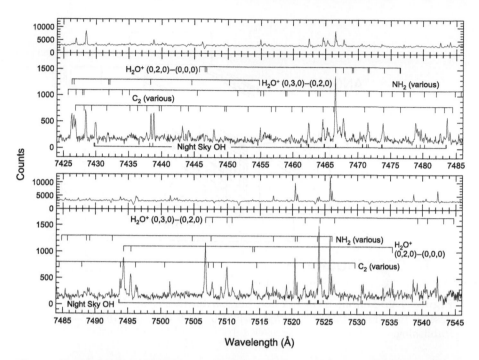

Figure 7.1: Spectrum of Comet 153P/Ikeya-Zhang, recorded 15,000 km tailward from the optocenter, illustrating a well-developed ion tail. The narrow panel for each half of the order is the optocenter spectrum, while the wide panel is the tail spectrum. All lines from two H_2O^+ bands are marked, along with lines of NH_2 and C_2 (Phillips). Not all the marked ionic lines are present; only the lower J-values appear in the spectrum. Comparison of the tail and the optocenter spectra show the increased strength of the ionic lines relative to the neutrals. The solar continuum has not been removed from either spectrum. This spectrum was obtained with the McDonald Observatory 2-D-coudé at R = 60,000. Reprinted from Feldman et al., *Comets II*, pp. 425–447, 2004 [6], by permission of the University of Arizona Press.

7.2.2 Experimental Spectrum of H_2O^+ Recorded by Hin Lew at NRC Canada

Lew [19] analysed many bands of the $\tilde{A}\,^1A_1 - \tilde{X}\,^1B_1$ electronic emission spectrum of H_2O^+ that occur in the wavelength region 4000–7500 Å. These include bands that have been observed in the tails of comets. The wavelengths and wave numbers of all assigned lines were tabulated; they are reproduced in the Appendix, Section 8.1. The assigned lines constitute only a fraction of the total number of lines in the laboratory spectra, but they do include most of the strong bands. Some predicted microwave and infrared lines that may be of astrophysical interest were included. Accurate rotational constants for the first three bending vibrational levels of the ground state were given, as well as energy levels in the upper and lower electronic states. The O-H bond distance

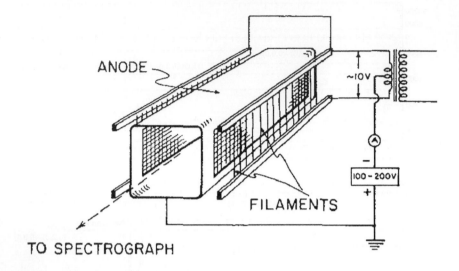

Figure 7.2: Multifilament electron bombardment source. Reproduced with permission from Lew, *Canad. J. Phys.*, 54:2028, 1976 [19]. © 1976 Canadian Science Publishing.

and the H-O-H angle in the $\tilde{X}(0,0,0)$ level were found to be 0.9988 Å and 110.46°, respectively.

The spectrum was first produced by the ion source illustrated in Figure 7.2. It was a diode of which the anode was a hollow square cylinder 5 cm × 5 cm × 15 cm made of stainless steel. On two opposing sides of the anode were windows covered with stainless steel mesh. Facing these windows were two banks of tungsten filaments 0.2 mm in diameter and 5 cm long spaced about 6 mm apart. These filaments were located about 3 mm from the mesh windows, and brought to thermionic emission temperatures. Electrons were accelerated from the filaments into the anode chamber by means of a potential of about 100 V. The total emission current was of the order of 0.25 A. This arrangement was housed in a vacuum tank, pumped by a 10 cm diffusion pump. The H_2O^+ spectrum was first observed when water vapour was let into the chamber to a pressure of 6×10^{-4} Torr. With the aid of multiple-traversal mirrors on the axis of the ion source, observations were first made photoelectrically on a 1 m monochromator. This was followed by photographic measurements on a 6 m Eagle-type spectrograph. Only emission spectra were seen.

A difficulty with the source shown in Figure 7.2 was the tendency for arcing to occur between the cathode and anode. Consequently, a simpler ion source was adopted, as shown in Figure 7.3. The anode is a water-cooled aluminium plate with grooves cut to reduce reflections. The cathode is a single bank of tungsten filaments set in slots in another water-cooled aluminium plate.

Because of the weakness of the emission, multiple-traversal mirrors

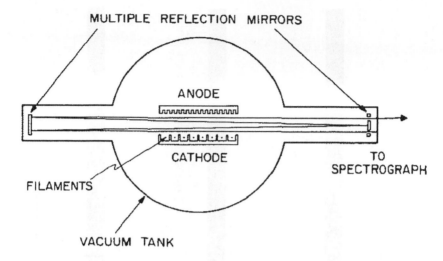

Figure 7.3: Multifilament hot cathode discharge source. Reproduced with permission from Lew, *Canad. J. Phys.*, 54:2028, 1976 [19]. © 1976 Canadian Science Publishing.

(a White cell) of 1 m radius of curvature was used on the source to give greater optical depth. Although the mirrors were set for 28 traversals, the actual optical depth was less than $29 \times 25 = 700$ cm, because of the 90% maximum reflection of the aluminized mirrors. The physical separation of 1 m reduced the amount of direct radiation that could go through the spectrograph.

The spectrum was photographed with a 6.5 m Eagle-type spectrograph that was air mounted. Two gratings were used, a 1200 line/mm for wavelengths below 7000 Å, and a 600 line one for wavelengths above 7000 Å. The respective linear dispersions were 1.2 Å/mm and 2.4 Å/mm, and 1$^\text{st}$ order resolving powers of 175,000 and 90,000 were attained.

Spectra in the range 3400–7000 Å were taken on Kodak type 103aF plates, and exposure times were from $\frac{1}{2}$ to $1\frac{1}{2}$ h. In the region 7000–9000 Å, Kodak type 1-N plates were used after hypersensitization. The exposure required was about 11 h taken over two or more days. A single exposure had to be spread over several days because the tungsten filament usually lasted only 3 to 4 h. The slit width was 0.030 mm throughout the measurement sequence. The spectrographic plates were measured using an automatic measuring machine developed for the laboratory by J. W. C. Johns [23].

The spectrum is quite extensive, with the strongest bands appearing in the 5500–6500 Å region. Some of the measured subbands are shown in Figures 7.4 and 7.5. Numbering of the vibrational levels of the upper state is that of a linear molecule. In this scheme, even values of v_2' give rise to $\Pi, \Phi, \ldots$ subbands, while odd values of v_2' give rise to $\Sigma, \Delta, \ldots$ subbands.

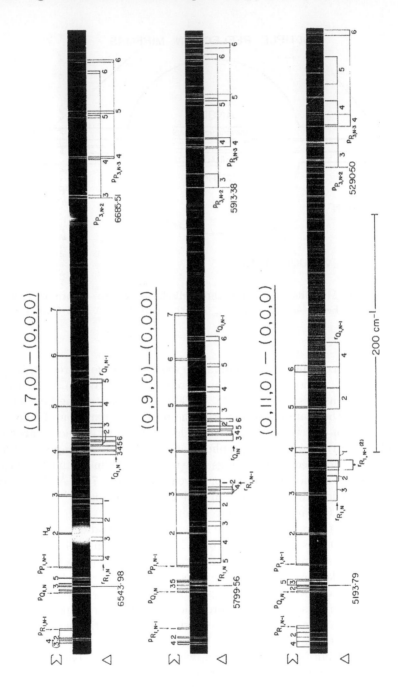

Figure 7.4: Σ and Δ subbands of (7–0), (9–0), and (11–0). Reproduced with permission from Lew, *Canad. J. Phys.*, 54:2028, 1976 [19]. © 1976 Canadian Science Publishing.

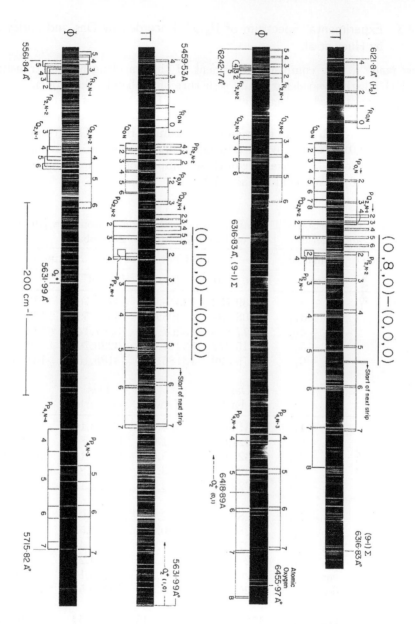

Figure 7.5: Π and Φ subbands of (8–0) and (10–0). Reproduced with permission from Lew, *Canad. J. Phys.*, 54:2028, 1976 [19]. © 1976 Canadian Science Publishing.

In 1987, Lew and Groleau [24] still recorded the electronic spectrum of the deuterated water cation, D_2O^+. Their tables of the assigned lines are also reproduced in the Appendix, Section 8.2.

7.2.3 Experimental Spectrum of H_2O^+ Recorded by Das and Farley and by Huet et al.

Lew made a very complete assessment of the wavelengths and transitions of H_2O^+, listing wavelengths and structure of subbands. Subsequently, other groups have studied a selection of subbands of H_2O^+. One of the first examples of this was the study of line centres and intensities by Das and Farley [25], and by Huet et al. [26] in the visible spectrum of H_2O^+. They compared the $(0,7,0)–(0,0,0)$ spectrum recorded by Lew [19] to the visible spectrum of H_2O^+ that they recorded. An example of these comparisons is shown for part of the $(0,7,0)–(0,0,0)$ Σ subband in Table 7.1.

Table 7.1: Comparison of Transition Frequencies, In cm^{-1}, from Lew [19], Das and Farley [25], and Huet et al. [26] For The $(0,7,0)–(0,0,0)$ Σ Subband

Branch	Transition	$F_1 - F_1$ Line Center		$F_2 - F_2$ Line Center	
		Huet et al.	Lew	Huet et al.	Lew
$^PP_{1,N-1}$	$1_{01} - 2_{11}$	13339.752	13339.723	13337.783	13337.742
	$2_{02} - 3_{12}$	13308.431	13305.096	13304.597	13304.600
	$3_{03} - 4_{13}$	13277.859	13267.127	13267.143	13267.127
	$4_{04} - 5_{14}$	13215.277	13227.070	13226.794	13226.778
	$5_{05} - 6_{15}$	13179.829	13185.105	13185.154	13185.105
		Das and Farley	Lew	Das and Farley	Lew
$^PR_{1,N-1}$	$2_{02} - 1_{10}$	15314.60	15314.594	15313.16	15313.173
	$3_{03} - 2_{11}$	15321.63	15321.621	15320.85	15320.868
	$4_{04} - 3_{12}$	15325.64	15325.641	15325.10	15325.106
	$5_{05} - 4_{13}$	15323.44	15323.442	15322.84	15322.831
	$6_{06} - 5_{14}$	15322.46	16322.457	15322.18	15322.174
$^PQ_{1,N}$	$1_{01} - 1_{11}$	15282.85	15282.846	15281.94	15281.982
	$2_{02} - 2_{12}$	15280.62	15280.628	15279.58	15279.589
	$3_{03} - 3_{13}$	15276.99	17277.007	15276.44	15276.459
	$4_{04} - 4_{14}$	15274.55	15274.559	15274.25	15274.267
	$5_{05} - 5_{15}$	15269.21	15269.218	15270.10	15270.105
	$6_{06} - 6_{16}$	15269.55	15269.569	15269.55	15269.569
$^PP_{1,N-1}$	$0_{00} - 1_{10}$	15261.33	15261.345	15260.44	15260.435
	$2_{02} - 3_{12}$	15200.84	15200.837	15199.68	15199.697
	$3_{03} - 4_{13}$	15163.89	15163.894	15163.12	15163.127
	$4_{04} - 5_{14}$	15125.62	15125.607	15125.08	15125.074
	$5_{05} - 6_{15}$	15082.80	15082.801	15083.29	15083.300
	$6_{06} - 7_{16}$	15045.19	15045.205	15044.79	15044.794

The $(0,7,0)–(0,0,0)$ Σ subband was labelled $(0,5,0)–(0,0,0)$ by Lew.

With the detection of H_2O^+ in comet tails through its optical spectrum [5, 27, 28] and the observation of N_2H^+ and HCO^+ in space through their microwave spectra [29–32], there was considerable interest among astrophysicists to see if H_2O^+ could be detected in space through its microwave or infrared spectrum. Both the microwave and ν_2 infrared spectra would be of type B with the selection rules:

$$\Delta J = 0, \pm 1; \ \ \Delta N = 0, \pm 1, \pm 2 \tag{7.1}$$

$$\Delta K_a = \pm 1, \pm 3 \cdots ; \ \ \Delta K_c = \pm 1, \pm 3 \cdots \tag{7.2}$$

with the addition of $\Delta v_1 = \pm 1$ for the infrared lines.

7.3   THE HERSCHEL SPACE OBSERVATORY

With the resultant invention of Herschel, a European Space Agency (ESA) space observatory [33], the first science highlights became plausible. The Heterodyne Instrument for the Far-Infrared (HIFI), is one of the three instruments on board. Herschel was launched in May 2009, and first results were published in the following year in two special features issues of *Astronomy & Astrophysics*: Vol. 518 (Herschel: the first science highlights) and Vol. 521 (Herschel/HIFI: first science highlights). Some examples, for H_3O^+, H_2O^+, OH^+, H_2O, and SiC_2, are presented below.

7.3.1   The Herschel-Heterodyne Instrument for the Far-Infrared (HIFI)

Th. de Graauw, F. P. Helmich, T. G. Phillips, J. Stutzki and colleagues [34].

HIFI, the Heterodyne Instrument for the Far-Infrared, was designed to provide very high spectral resolution over the widest possible frequency range. As the 3.5 m telescope has a limited collecting area, near quantum limited noise temperatures are required for sensitivity. Owing to the latest developments in correlator and acousto-optical spectrometer (AOS) spectrometer technology, it was possible to construct an instrument with the following capabilities:

1. Continuous frequency coverage from 480–1250 GHz in five bands, and from 1410–1910 GHz in two bands.

2. Spectral resolutions between 300 and 0.03 $km\,s^{-1}$.

3. Detection sensitivity close to the fundamental quantum noise limit.

In Figure 7.6, a block diagram of HIFI is given in which subsystems and their interconnections are shown. These are:

1. The focal plane subsystem with the focal plane unit (FPU) mounted on the optical bench on top of the liquid He vessel inside the cryostat. A focal plane control unit (FCU) is located at the service module (SVM).

This supplies the bias voltage for mixers and intermediate frequency (IF) preamplifiers in the focal plane subsystem. It also controls the local oscillator (LO) diplexers, the focal plane chopper mechanism and the calibration source.

2. The LO subsystem with the local oscillator unit (LCU) located on the outside wall of the Herschel cryostat. It accomodates 7 local oscillator assemblies (LOA), each of which containing two local oscillator chains.

3. The wide-band spectrometer (WBS). It is composed of a pair of array AOSs which offer, for each of the two polarisations, a frequency resolution of about 1 MHz and a bandwidth of 4 GHz.

4. The high-resolution spectrometer (HRS). It consists of a pair of auto-correlator spectrometers that are divided into subbands with several combinations of bandwidth and frequency resolution. Each subband can be placed anywhere within the full 4 GHz IF band.

5. The instrument control unit (ICU). Mounted on the service module, it receives the commands that control the operation of the instrument and returns science and housekeeping data.

Figures 7.7 and 7.8 show the common optics assembly layout (Figure 7.7) and the flight model focal plane unit (Figure 7.8).

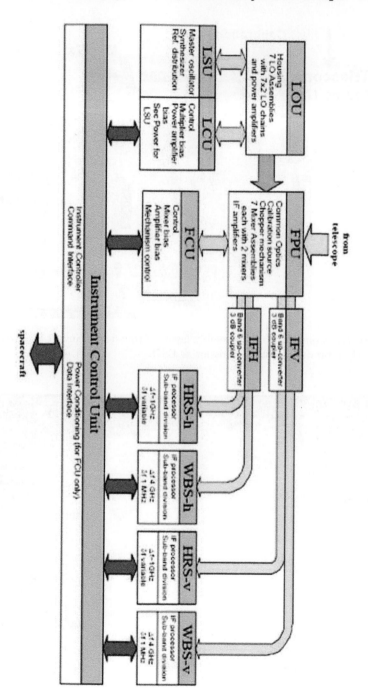

Figure 7.6: HIFI block diagram. From de Graauw et al., *A&A*, 518:L6, 2010 [34], reproduced with permission © ESO.

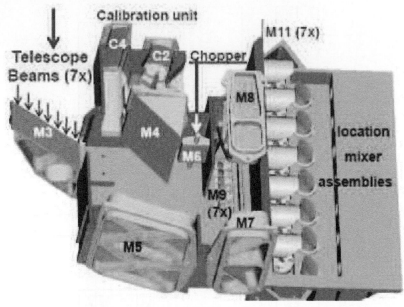

Figure 7.7: Common optics assembly layout. From de Graauw et al., *A&A*, 518:L6, 2010 [34], reproduced with permission © ESO.

Figure 7.8: The flight model focal plane unit. From de Graauw et al., *A&A*, 518:L6, 2010 [34], reproduced with permission © ESO.

Figure 7.9 shows the in-orbit system noise temperatures for the 5 superconductor-insulator-superconductor (SIS) and the two hot electron bolometer (HEB) bands for both linear polarizations. The data were taken with the WBS spectrometers.

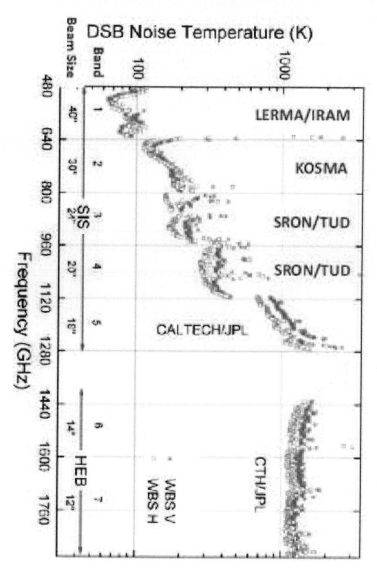

Figure 7.9: HIFI's T_{sys} double side band (DSB) for H and V polarization for the 7 bands, together with their aimed angular resolution half power beam width (HPBW). Also shown are the institutes that provided the mixers. From de Graauw et al., *A&A*, 518:L6, 2010 [34], reproduced with permission © ESO.

7.3.2 Detection of Interstellar Oxidaniumyl: Abundant H_2O^+ Towards the Star-Forming Regions DR21, Sgr B2, and NGC 6334

V. Ossenkopf, H. S. P. Müller, D. C. Lis, and P. Schilke and colleagues [35].

The H_2O^+ cation, oxidaniumyl, has bond lengths and angle slightly larger than H_2O. Consequently, the ground state dipole moment is slightly smaller than that of water, 2.4 D, according to quantum-chemical calculations [36]. Since the electronic ground state is $^2B_1$, ortho and para levels are reversed relative to water. Figure 7.10 shows a diagram of the lowest rotational levels of ortho-H_2O^+.

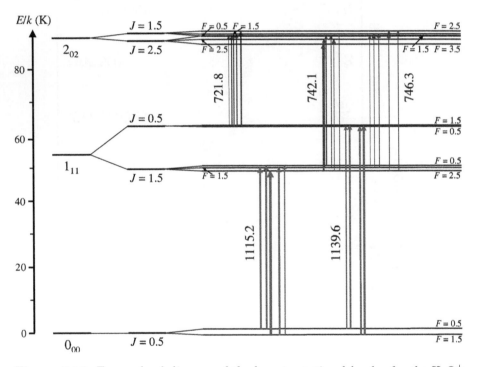

Figure 7.10: Energy level diagram of the lowest rotational levels of ortho-H_2O^+ and its radiative transitions. The fine structure transition frequencies are given in GHz. From Ossenkopf et al., *A&A*, 518:L111, 2010 [35], reproduced with permission © ESO.

The rotational spectrum was measured by laser magnetic resonance spectroscopy by Strahan et al. [37] and by Mürtz et al. [38]. When comparing the predictions of the $N_{K_aK_c} = 1_{11} - 0_{00}$, $J = 3/2 - 1/2$ fine structure component near 1115 GHz made from the parameters derived from the two experiments, Table 7.2, a discrepancy of between 27.3 and 28.5 MHz is found, even though both articles claim to have reproduced the experimental data to ~ 2 MHz. That such an accuracy is in principle achievable has been demonstrated by Brown and Müller [39], who reanalysed the equivalent measurements of SH^+.

Table 7.2: Parameters of Hyperfine Lines $F' - F''$ of Observed Ortho-H_2O^+ transitions, Including Predicted Frequencies, Einstein-A and Optical Depth at Low Temperatures.

Transition upper – lower	$\nu^a_{M\ddot{u}rtz}$ [MHz]	$\nu^a_{Strahan}$ [MHz]	$\nu^b_{OHbased}$ [MHz]	A [s^{-1}]	$\int \tau d\nu$
o-H_2O^+ $1_{11} - 0_{00}$ $J = 3/2 - 1/2$					
$F = 5/2 - 3/2$	1115204.1	1115175.8	1115161	0.031	23.51
$3/2 - 1/2$	1115150.5	1115122.0	1115107	0.017	8.67
$3/2 - 3/2$	1115263.2	1115235.6	1115221	0.014	7.00
$1/2 - 1/2$	1115186.2	1115158.0	1115143	0.027	6.96
$1/2 - 3/2$	1115298.9	1115271.6	1115257	0.0035	0.88

[a]Predictions based on Strahan et al. [37] and Mürtz et al. [38]. Nominal uncertainties are 2 MHz. However, this is inconsistent with the discrepancy between the two predictions so that the actual uncertainty is unknown. [b]From the matching DR21 OH pattern by Guilloteau et al. [40].

The spectra seen in the different regions are shown in Figures 7.11 to 7.13. Figure 7.14 compares the H_2O^+ velocity profile with other species found in the same region.

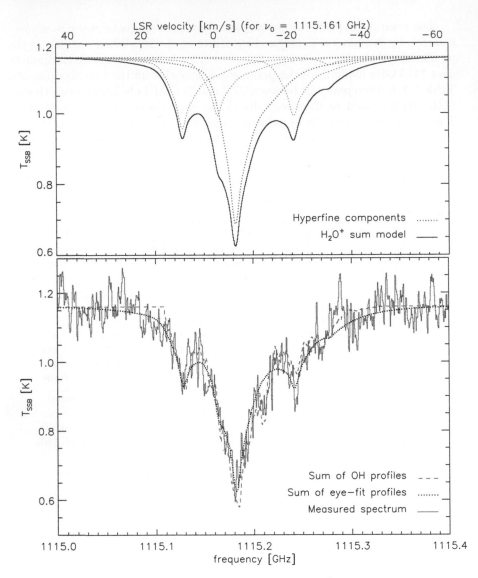

Figure 7.11: Fit of the hyperfine multiplet of the H_2O^+ 1115 GHz line in DR21. The bottom panel shows the 0.5 K absorption line superimposed on two different fit profiles, one based on a 3-component Gaussian and the other one using the OH 6 cm absorption spectrum from Guilloteau et al. [40]. The top panel shows a breakdown of the fitted profile into its hyperfine constituents in the case of the 3-Gaussian profile. From Ossenkopf et al., *A&A*, 518:L111, 2010 [35], reproduced with permission © ESO.

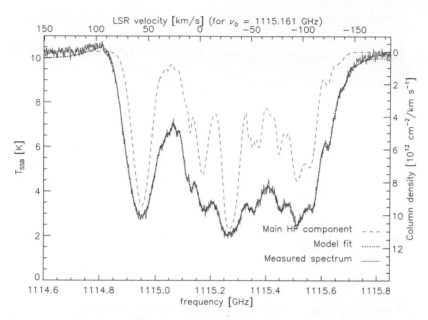

Figure 7.12: Fit of the observed H_2O^+ line in Sgr B2. The dashed line visualises the velocity structure of the absorbers by plotting the strongest hyperfine component on a linear column density scale, i.e., without optical depth correction. From Ossenkopf et al., *A&A*, 518:L111, 2010 [35], reproduced with permission © ESO.

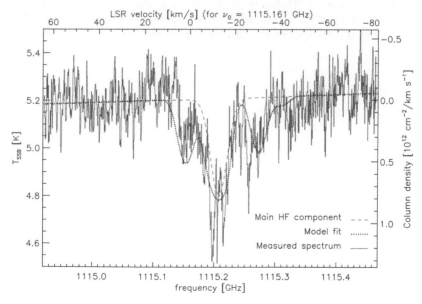

Figure 7.13: Same as Figure 7.12, but for NGC 6334. From Ossenkopf et al., *A&A*, 518:L111, 2010 [35], reproduced with permission © ESO.

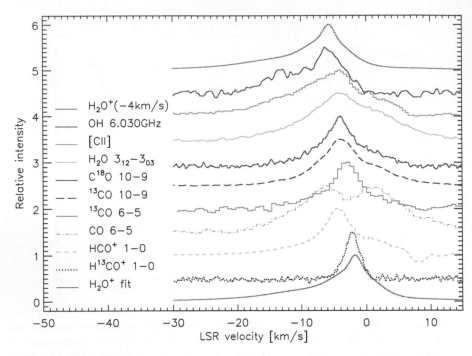

Figure 7.14: Comparison of the fitted H_2O^+ velocity profile to other tracers observed in DR21 with similar beam size. The profiles are normalised to a peak of unity and separated by multiples of 0.5 from bottom to top. The fit (bottom line) used the Strahan et al. [37] based line frequency prediction, the profile at the top is shifted by -4.0 km s^{-1}, corresponding to a rest frequency lower by 15 MHz. From Ossenkopf et al., *A&A*, 518:L111, 2010 [35], reproduced with permission © ESO.

7.3.3 Interstellar OH^+, H_2O^+, and H_3O^+ Along the Sight-Line to G10.6–0.4

M. Gerin, M. De Luca, J. Black, J. R. Goicoechea, E. Herbst and colleagues [41].

Rotational transitions of OH^+ ($\tilde{X}\,^3\Sigma^-$), H_2O^+ ($\tilde{X}\,^2B_1$), and H_3O^+ ($\tilde{X}\,^1A_1$) have been observed towards the submillimeter continuum source G10.6-0.4 (W31C). These are shown in Table 7.3. The spectra, presented in Figures 7.15 and 7.16, show deep absorption over a broad velocity range that originates in the interstellar matter along the line of sight as well as the molecular gas directly associated with this source.

Table 7.3: Transition spectroscopic parameters of OH^+, H_2O^+ and H_3O^+

Transition upper – lower	Frequency [MHz]	Error [MHz]	E_l [cm^{-1}]	A [10^{-2}s^{-1}]	Ref.
\multicolumn{6}{c}{OH^+ $N = 1 - 0$}					
\multicolumn{6}{c}{$J = 2 - 1$}					
$F = 5/2 - 3/2$	971803.8	1.5	0.0	1.82	1
$3/2 - 1/2$	971805.3	1.5	0.0	1.52	1
$3/2 - 3/2$	971919.2	1.0	0.0	0.30	1
\multicolumn{6}{c}{o-H_2O^+ $1_{11} - 0_{00}$}					
\multicolumn{6}{c}{$J = 3/2 - 1/2$}					
$F = 3/2 - 1/2$	1115122.0	10	0.0	1.71	2
$1/2 - 1/2$	1115158.0	10	0.0	2.75	2
$5/2 - 3/2$	1115175.8	10	0.0	3.10	2
$3/2 - 3/2$	1115235.6	10	0.0	1.39	2
$1/2 - 3/2$	1115271.6	10	0.0	0.35	2
\multicolumn{6}{c}{p-H_2O^+ $1_{10} - 1_{01}$}					
\multicolumn{6}{c}{$J = 3/2 - 3/2$}					
$F = 3/2 - 3/2$	607207.0	20	20.9	0.60	2
\multicolumn{6}{c}{H_3O^+}					
$0_0^- - 1_0^+$	984711.9	0.3	5.1	2.3	3

1. Reference [42] & CDMS; 2. Reference [35,37]; 3. Reference [43] and Jet Propulsion Laboratory (JPL) catalog.

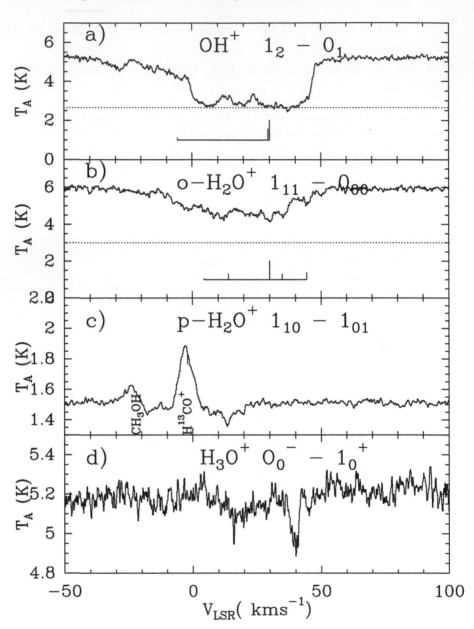

Figure 7.15: HIFI spectra of a) OH^+; b) $o\text{-}H_2O^+$; c) $p\text{-}H_2O^+$; and d) H_3O^+. The hyperfine structure of the OH^+ and $o\text{-}H_2O^+$ lines (see Table 7.3) is indicated. In both cases, the velocity scale corresponds to the rest frequency of the strongest hyperfine component. In panel c), the velocity scale refers to the frequency of the $H^{13}CO^+$ (6–5) line at 607174.65 MHz. Panel d) uses the frequency of 984711.907 MHz. From Gerin et al., *A&A*, 518:L110, 2010 [41], reproduced with permission © ESO.

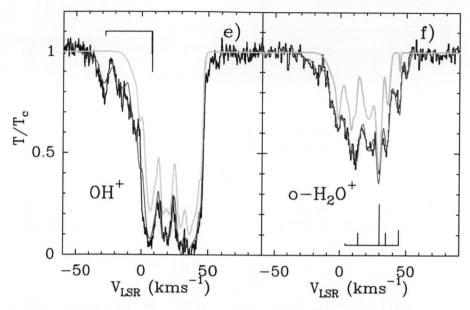

Figure 7.16: Continuation of Figure 7.15. Panels e) and f) show overlays with the synthetic spectra constructed with empirical models. Grey curves show the profile of the strongest hyperfine component. From Gerin et al., *A&A*, 518:L110, 2010 [41], reproduced with permission © ESO.

7.3.4 Detection of OH^+ and H_2O^+ Towards Orion KL

H. Gupta, P. Rimmer, J. C. Pearson, S. Yu, and E. Herbst and colleagues [44].

A number of transitions of OH^+, H_2O^+, and H_3O^+ have been observed towards the star-forming region Orion KL, as detailed in Table 7.4.

Table 7.4: Spectroscopic Parameters of the Observed Transitions

Transition upper – lower	Frequency [MHz]	Error [MHz]	E_l [cm^{-1}]	g_l	g_u	$\mu^2 S^a$ [D^2]	$10^2 A_{ij}$ [s^{-1}]
		OH^{+b} $N = 1 - 0$					
		$J = 1 - 1$					
$F = 1/2 - 1/2$	909045.2	1.5	0.004	4	6	1.20	1.05
$1/2 - 3/2$	909158.8	1.5	0	2	4	2.40	0.52
		$J = 2 - 1$					
$F = 5/2 - 3/2$	971803.8	1.5	0	4	6	10.24	1.82
$3/2 - 1/2$	971805.3	1.5	0.004	2	4	5.69	1.52
$3/2 - 3/2$	971919.2	1.0	0	4	4	1.14	0.30
		o-H$_2$O^{+c} $1_{10} - 0_{00}$					
		$J = 3/2 - 1/2$					
$F = 3/2 - 1/2$	1115150.0	1.8	0.004	2	4	4.14	1.67
$1/2 - 1/2$	1115186.0	1.8	0.004	2	2	3.32	2.68
$5/2 - 3/2$	1115204.0	1.8	0	4	6	11.23	3.02
$3/2 - 3/2$	1115263.0	1.8	0	4	4	3.35	1.35
$1/2 - 3/2$	1115298.7	1.8	0	2	2	0.42	0.34
		$J = 1/2 - 1/2$					
$F = 3/2 - 1/2$	1139541.1	1.8	0.004	2	2	0.42	3.61
$1/2 - 1/2$	1139560.6	1.8	0.004	2	4	3.35	1.44
$5/2 - 3/2$	1139653.5	1.8	0	4	2	3.32	2.86
$3/2 - 3/2$	1139673.3	1.8	0	4	4	4.14	1.78
		H$_3$O^{+d}					
$0_0^- - 1_0^+$	984711.9	0.1^e	5.1	4	12	8.30	2.30
$1_1^- - 1_1^+$	1655834.8	0.3^f	0	6	6	6.22	5.48
$2_2^- - 2_2^+$	1657248.4	0.3^e	29.6	10	10	4.67	7.32

[a] Dipole moments (μ): 2.256 D (OH$^+$; Werner et al. [45]); 2.37 D (H$_2$O$^+$; Wu et al. [46]); 1.44 D (H$_3$O$^+$; Botschwina et al. [47]). Frequencies from: [b] Müller et al. [42]; [c] Mürtz et al. [38]; [d] Yu et al. [43]. [e] $\int T_A dv < 0.482$ K km s^{-1} for the 984.7 GHz line, and < 2.412 K km s^{-1} for the 1657.2 GHz line. [f] Blended with a strong $2_{12} - 1_{10}$ ortho-H$_2^{18}$O line at 1655831 MHz.

7.3.5 Water Production In Comet 81P/Wild 2 as Determined by Herschel/HIFI

M. de Val-Borro, P. Hartogh, J. Crovisier, D. Bockelée-Morvan, and colleagues [48].

The fundamental ortho- and para-water rotational transitions at 556.936 GHz and 1113.343 GHz were detected using the high resolution spectrometer (HRS). Standard dual beam switch (DBS) cross-maps were obtained to study the excitation conditions throughout the coma. These are shown in Figures 7.17 to 7.20. Figure 7.21 shows the intensities of the $1_{10} - 1_{01}$ and $1_{11} - 0_{00}$ lines at different times. The water production rate was calculated to be in the range $0.9 - 1.1 \times 10^{28}\,\mathrm{s}^{-1}$, using a radiative transfer code that includes collisional effects and infrared fluorescence by solar radiation. The water production rates over time are shown in Figure 7.22.

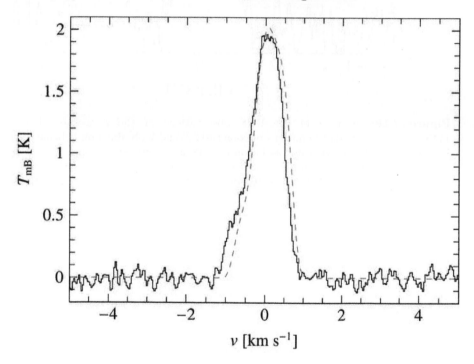

Figure 7.17: Central point in the DBS map of the ortho-water line at 556.936 GHz in comet 81P/Wild 2 obtained by the HRS on February 4.24 UT. The velocity scale is given with respect to the comet rest frame with a resolution of $\sim 32\,\mathrm{m\,s}^{-1}$. A synthetic line profile for isotropic outgassing with $v_{\exp} = 0.6\,\mathrm{km\,s}^{-1}$, $T = 40$ K, and $x_{n_e} = 0.2$ is shown by the dashed line. From de Val-Borro et al., *A&A*, 521:L50, 2010 [48], reproduced with permission © ESO.

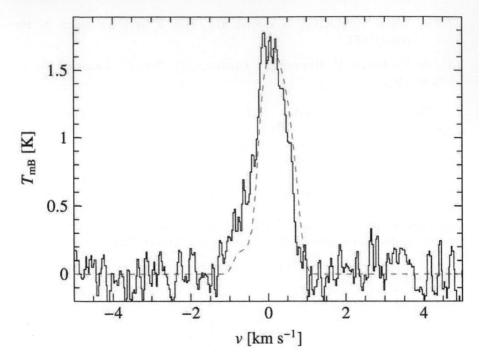

Figure 7.18: Average of the HRS observations of the para-water line at 1113.343 GHz towards the nucleus of comet 81P/Wild 2 obtained on February 2.13 UT and 2.15 UT. The velocity resolution is $\sim 32 \, \text{m s}^{-1}$ after smoothing. The dashed line shows the best fit profile for the same parameters as in Figure 7.17. From de Val-Borro et al., *A&A*, 521:L50, 2010 [48], reproduced with permission © ESO.

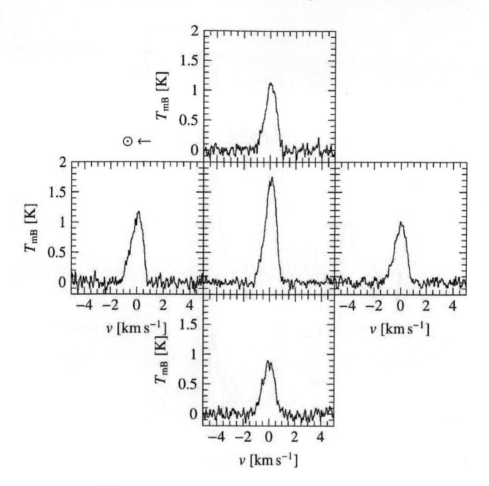

Figure 7.19: DBS cross-map of the ortho-water $1_{10} - 1_{01}$ transition obtained with the HRS on February 1.44 UT. Offset positions are $\sim 22''$ apart. The velocity resolution is ~ 32 m s^{-1}. The sun direction is indicated. From de Val-Borro et al., *A&A*, 521:L50, 2010 [48], reproduced with permission © ESO.

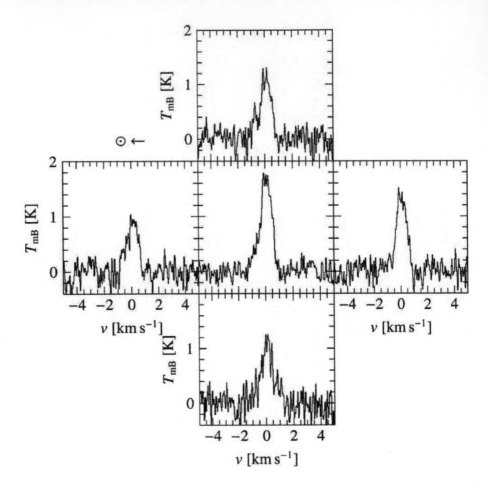

Figure 7.20: Average of the cross-maps of the $1_{11} - 0_{00}$ para-water line obtained with the HRS on February 2.13 and 2.15 UT. The offset spacing is $\sim 10''$ and the effective velocity resolution after smoothing is ~ 32 m s^{-1} . The sun direction is indicated. From de Val-Borro et al., *A&A*, 521:L50, 2010 [48], reproduced with permission © ESO.

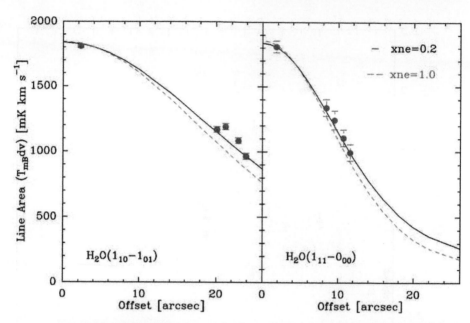

Figure 7.21: Intensities of the $1_{10} - 1_{01}$ (Feb. 1.44 UT, left panel) and $1_{11} - 0_{00}$ (Feb. 2.13–2.15 UT, right panel) lines as a function of position offset from the estimated position of the peak brightness. Model calculations for $x_{n_e} = 0.2$ and 1 are shown with a black solid line and red dashed line, respectively. We assumed T = 40 K and $v_{exp} = 0.6\,\mathrm{km\,s^{-1}}$. From de Val-Borro et al., $A\&A$, 521:L50, 2010 [48], reproduced with permission © ESO.

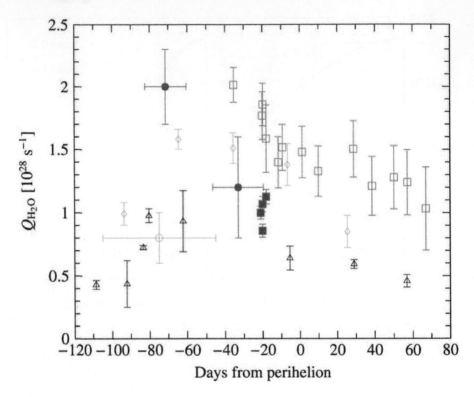

Figure 7.22: Water production rates with 1-σ uncertainties in comet 81P/Wild 2 as a function of time from perihelion. HIFI measurements are shown by squares. Filled circles represent OH 18-cm line observations with the Nançay radio telescope. Water production rates for the 1997 apparition from Crovisier et al. [49], Farnham and Schleicher [50], Fink et al. [51], and Mäkinen et al. [52] are shown by empty circles, triangles, diamonds, and squares, respectively. From de Val-Borro et al., *A&A*, 521:L50, 2010 [48], reproduced with permission © ESO.

7.3.6 A High-Resolution Line Survey of IRC +10216 with Herschel/HIFI – First Results: Detection of Warm Silicon Dicarbide

J. Cernicharo, L. B. F. M. Waters, L. Decin, P. Encrenaz, and colleagues [53].

Cernicharo and colleagues [53] used the Herschel/HIFI instruments to study the evolved star IRC+10216 and gain further understanding of the complex chemistry of the circumstellar envelopes (CSEs). This instrument gives wide coverage and also high spectral resolution, allowing the complete list of the chemical content and details of the inner and outer zones of the CSEs. The full spectrum is shown in Figure 7.23, with some detailed spectra over 3 GHz wide range.

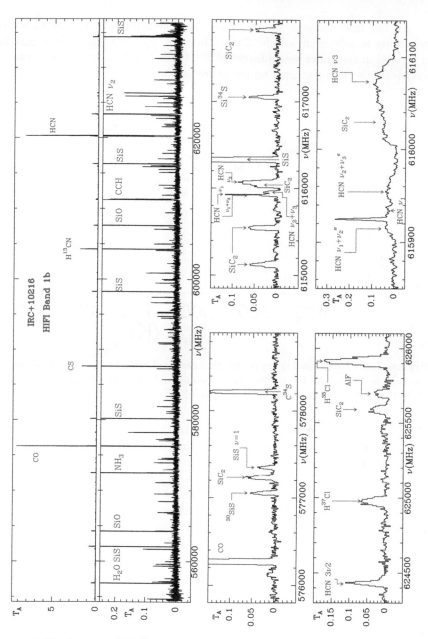

Figure 7.23: Spectra of IRC +10216 observed with HIFI band 1b. The two upper panels present the complete spectrum on two different intensity scales. The panels below show different 3 GHz wide ranges of the survey. All data have been smoothed to a spectral resolution of 2.8 km s^{-1} except for the right bottom panel, which shows the spectrum around several vibrational lines of HCN with the nominal WBS resolution (1.1 MHz, $\simeq$0.5 km s^{-1}). From Cernicharo et al., *A&A*, 521:L8, 2010 [53], reproduced with permission © ESO.

All the lines, except for one, have been identified, including 41 features belonging to silicon dicarbide, SiC_2. The energy of the lines shows that SiC_2 is formed in the inner warm layers of the CSE. The line intensities were used to determine the rotational temperature as shown in Figure 7.24.

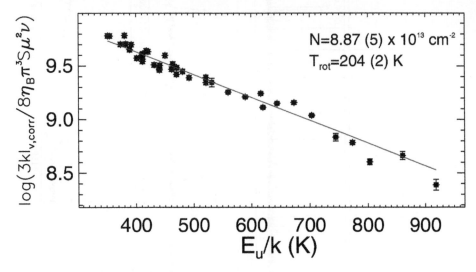

Figure 7.24: Rotational diagram for the observed SiC_2 lines. The rotational temperature is 204 K, and the beam averaged column density of SiC_2 is 8.9×10^{13} cm^{-2}. From Cernicharo et al., *A&A*, 521:L8, 2010 [53], reproduced with permission © ESO.

Cernicharo and colleagues described, in their paper [53], a model for the abundance of SiC_2 in the inner envelope and expansion to the outer envelope. This is shown in Figure 7.25. The abundance of SiC_2 in the outer envelope is enhanced mainly by the reaction between Si and C_2H_2. There is also a contribution from the reaction between Si^+ and C_2H. Their analysis contributed to the understanding of the chemistry involved in the interaction between the CSEs. Figure 7.26 shows the observed lines and the calculated profiles for the SiC_2 in the inner envelope.

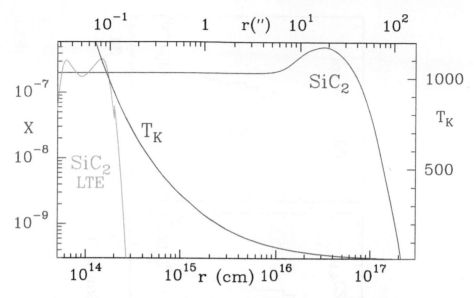

Figure 7.25: Abundance of SiC_2, X, derived from the chemical model described fully in Reference [53] (red line) and in thermodynamical equilibrium (green line). The blue line shows the kinetic temperature, T_K, of the gas. The axis shows the distance to the star in cm (bottom) and the angular distance (top) as seen from the Earth ($d = 120$ pc). From Cernicharo et al., *A&A*, 521:L8, 2010 [53], reproduced with permission © ESO.

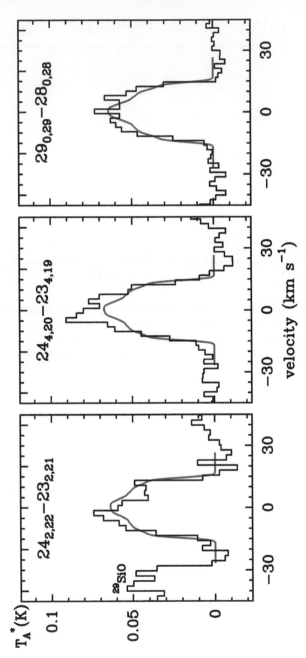

Figure 7.26: Comparison between the line profiles of three rotational transitions of SiC$_2$, as observed with HIFI (black histograms) and as calculated with the radiative transfer model (red lines) using the abundance profile shown in red in Figure 7.25. From Cernicharo et al., *A&A*, 521:L8, 2010 [53], reproduced with permission © ESO.

7.4 Bibliography

[1] H. Lew and I. Heiber. Spectrum of H_2O^+. *The Journal of Chemical Physics*, 58(3):1246–1247, 1973.

[2] F. L. Whipple. A comet model. I. The acceleration of Comet Encke. *The Astrophysical Journal*, 111:375–394, March 1950.

[3] F. L. Whipple. A comet model. II. Physical relations for comets and meteors. *The Astrophysical Journal*, 113:464–474, May 1951.

[4] G. Herzberg and H. Lew. Tentative identification of the H_2O^+ ion in comet Kohoutek. *Astronomy and Astrophysics*, 31:123–124, February 1974.

[5] P. A. Wehinger, S. Wyckoff, G. H. Herbig, G. Herzberg, and H. Lew. Identification of H_2O in the tail of comet Kohoutek (1973f). *The Astrophysical Journal*, 190:L43–L46, May 1974.

[6] P. D. Feldman, A. L. Cochran, and M. R. Combi. Spectroscopic investigations of fragment species in the coma. In M. C. Festou, H. U. Keller, and H. A. Weaver, editors, *Comets II*, pages 425–447. University of Arizona Press, Tucson, 2004.

[7] P. Swings. Cometary spectra (George Darwin Lecture). *Quarterly Journal of the Royal Astronomical Society*, 6:28–69, March 1965.

[8] P. Swings and L. Haser. *Atlas of Representative Cometary Spectra*. Liege Astrophysical Institute, 1956.

[9] C. Arpigny. Spectra of comets and their interpretation. *Annual Review of Astronomy & Astrophysics*, 3:351–376, 1965.

[10] L. L. Wilkening and M. S. Matthews. *Comets*. University of Arizona Press, Tucson, 1982.

[11] M. F. A'Hearn. Spectrophotometry of comets at optical wavelengths. In L. L. Wilkening and M. S. Matthews, editors, *Comets*, pages 433–460. University of Arizona Press, Tucson, 1982.

[12] M. C. H. Wright, I. de Pater, J. R. Forster, P. Palmer, L. E. Snyder, J. M. Veal, M. F. A'Hearn, L. M. Woodney, W. M. Jackson, Y.-J. Kuan, and A. J. Lovell. Mosaicked images and spectra of $J = 1 \to 0$ HCN and HCO^+ emission from comet Hale-Bopp (1995 o1). *The Astronomical Journal*, 116(6):3018–3028, 1998.

[13] D. C. Lis, D. M. Mehringer, D. Benford, M. Gardner, T. G. Phillips, D. Bockelée-Morvan, N. Biver, P. Colom, J. Crovisier, D. Despois, and H. Rauer. New molecular species in comet C/1995 O1 (Hale-Bopp) observed with the Caltech submillimeter observatory. *Earth, Moon, and Planets*, 78(1–3):13–20, 1997.

[14] R. Umbach, K. Jockers, and E. H. Geyer. Spatial distribution of neutral and ionic constituents in comet P/Halley. *Astronomy and Astrophysics Supplement Series*, 127(3):479–495, 1998.

[15] W. M. Jackson and B. Donn. Photochemical effects in the production of cometary radicals and ions. *Icarus*, 8(1):270–280, 1968.

[16] R. Wegmann, K. Jockers, and T. Bonev. H_2O^+ ions in comets: Models

and observations. *Planetary and Space Science*, 47(6–7):745–763, 1999.

[17] B. L. Lutz, M. Womack, and R. M. Wagner. Ion abundances and implications for photochemistry in Comets Halley (1986 III) and Bradfield (1987 XXIX). *The Astrophysical Journal*, 407:402–411, April 1993.

[18] C. Arpigny. Physical chemistry of comets: Models, uncertainties, data needs. In I. Nenner and P. Bréchignac, editors, *Molecules and Grains In Space: 50th International Meeting of Physical Chemistry, Mont Sainte-Odile, France, September 1993*, AIP conference proceedings. American Institute of Physics, 1994.

[19] H. Lew. Electronic spectrum of H_2O^+. *Canadian Journal of Physics*, 54(20):2028–2049, 1976.

[20] A. L. Cochran and W. D. Cochran. A high spectral resolution atlas of comet 122P/de Vico. *Icarus*, 157(2):297–308, 2002.

[21] T. Bonev and K. Jockers. H_2O^+ ions in the inner plasma tail of comet Austin 1990 V. *Icarus*, 107(2):335–357, 1994.

[22] H. A. Weaver, P. D. Feldman, M. R. Combi, V. Krasnopolsky, C. M. Lisse, and D. E. Shemansky. A search for argon and O VI in three comets using the far ultraviolet spectroscopic explorer. *The Astrophysical Journal Letters*, 576(1):L95, 2002.

[23] J. W. C. Johns, A. R. W. McKellar, and D. Weitz. Wavelength measurements of $^{13}C^{16}O$ laser transitions. *Journal of Molecular Spectroscopy*, 51(3):539–545, 1974.

[24] H. Lew and R. Groleau. Electronic spectrum of D_2O^+. *Canadian Journal of Physics*, 65(7):739–752, 1987.

[25] B. Das and J. W. Farley. Observation of the visible absorption spectrum of H_2O^+. *The Journal of Chemical Physics*, 95(12):8809–8815, 1991.

[26] T. R. Huet, I. Hadj Bachir, J.-L. Destombes, and M. Vervloet. The $\tilde{A}^1A_1 - \tilde{X}^1B_1$ transition of H_2O^+ in the near infrared region. *The Journal of Chemical Physics*, 107(15):5645–5651, 1997.

[27] P. Wehinger and S. Wyckoff. H_2O^+ in spectra of comet Bradfield (1974b). *The Astrophysical Journal*, 192:L41–L42, August 1974.

[28] J. Bortle, S. O'Meara, P. MacKinnon, D. Somers, D. Welch, R. Dick, F. Lossing, S. Murrell, C. Knuckles, E. J. Reese, J. S. Neff, D. A. Ketelsen, V. V. Smith, G. H. Herbig, D. Duncan, D. Soderblom, F. Pilcher, P. Maley, and K. Ikeya. Comet West (1975n). *International Astronomical Union Circulars*, 2927:1, March 1976.

[29] B. E. Turner. U93.174 – A new interstellar line with quadrupole hyperfine splitting. *The Astrophysical Journal*, 193:L83–L87, October 1974.

[30] S. Green, J. A. Montgomery, Jr., and P. Thaddeus. Tentative identification of U93.174 as the molecular ion N_2H^+. *The Astrophysical Journal*, 193:L89–L91, October 1974.

[31] D. Buhl and L. E. Snyder. Unidentified interstellar microwave line. *Nature*, 228:267–269, 1970.

[32] R. C. Woods, T. A. Dixon, R. J. Saykally, and P. G. Szanto. Laboratory microwave spectrum of HCO^+. *Physical Review Letters*, 35:1269–1272,

November 1975.

[33] G. L. Pilbratt, J. R. Riedinger, T. Passvogel, G. Crone, D. Doyle, U. Gageur, A. M. Heras, C. Jewell, L. Metcalfe, S. Ott, and M. Schmidt. Herschel space observatory – An ESA facility for far-infrared and submillimetre astronomy. *Astronomy and Astrophysics*, 518:L1, 2010.

[34] Th. de Graauw, F. P. Helmich, T. G. Phillips, J. Stutzki, E. Caux, N. D. Whyborn, P. Dieleman, P. R. Roelfsema, H. Aarts, R. Assendorp, R. Bachiller, W. Baechtold, A. Barcia, D. A. Beintema, V. Belitsky, A. O. Benz, R. Bieber, A. Boogert, C. Borys, B. Bumble, P. Caïs, M. Caris, P. Cerulli-Irelli, G. Chattopadhyay, S. Cherednichenko, M. Ciechanowicz, O. Coeur-Joly, C. Comito, A. Cros, A. de Jonge, G. de Lange, B. Delforges, Y. Delorme, T. den Boggende, J.-M. Desbat, C. Diez-González, A. M. Di Giorgio, L. Dubbeldam, K. Edwards, M. Eggens, N. Erickson, J. Evers, M. Fich, T. Finn, B. Franke, T. Gaier, C. Gal, J. R. Gao, J.-D. Gallego, S. Gauffre, J. J. Gill, S. Glenz, H. Golstein, H. Goulooze, T. Gunsing, R. Güsten, P. Hartogh, W. A. Hatch, R. Higgins, E. C. Honingh, R. Huisman, B. D. Jackson, H. Jacobs, K. Jacobs, C. Jarchow, H. Javadi, W. Jellema, M. Justen, A. Karpov, C. Kasemann, J. Kawamura, G. Keizer, D. Kester, T. M. Klapwijk, T. Klein, E. Kollberg, J. Kooi, P.-P. Kooiman, B. Kopf, M. Krause, J.-M. Krieg, C. Kramer, B. Kruizenga, T. Kuhn, W. Laauwen, R. Lai, B. Larsson, H. G. Leduc, C. Leinz, R. H. Lin, R. Liseau, G. S. Liu, A. Loose, I. López-Fernandez, S. Lord, W. Luinge, A. Marston, J. Martín-Pintado, A. Maestrini, F. W. Maiwald, C. McCoey, I. Mehdi, A. Megej, M. Melchior, L. Meinsma, H. Merkel, M. Michalska, C. Monstein, D. Moratschke, P. Morris, H. Muller, J. A. Murphy, A. Naber, E. Natale, W. Nowosielski, F. Nuzzolo, M. Olberg, M. Olbrich, R. Orfei, P. Orleanski, V. Ossenkopf, T. Peacock, J. C. Pearson, I. Peron, S. Phillip-May, L. Piazzo, P. Planesas, M. Rataj, L. Ravera, C. Risacher, M. Salez, L. A. Samoska, P. Saraceno, R. Schieder, E. Schlecht, F. Schlöder, F. Schmülling, M. Schultz, K. Schuster, O. Siebertz, H. Smit, R. Szczerba, R. Shipman, E. Steinmetz, J. A. Stern, M. Stokroos, R. Teipen, D. Teyssier, T. Tils, N. Trappe, C. van Baaren, B.-J. van Leeuwen, H. van de Stadt, H. Visser, K. J. Wildeman, C. K. Wafelbakker, J. S. Ward, P. Wesselius, W. Wild, S. Wulff, H.-J. Wunsch, X. Tielens, P. Zaal, H. Zirath, J. Zmuidzinas, and F. Zwart. The Herschel-Heterodyne Instrument for the Far-Infrared (HIFI). *Astronomy and Astrophysics*, 518:L6, 2010.

[35] V. Ossenkopf, H. S. P. Müller, D. C. Lis, P. Schilke, T. A. Bell, S. Bruderer, E. Bergin, C. Ceccarelli, C. Comito, J. Stutzki, A. Bacman, A. Baudry, A. O. Benz, M. Benedettini, O. Berne, G. Blake, A. Boogert, S. Bottinelli, F. Boulanger, S. Cabrit, P. Caselli, E. Caux, J. Cernicharo, C. Codella, A. Coutens, N. Crimier, N. R. Crockett, F. Daniel, K. Demyk, P. Dieleman, C. Dominik, M. L. Dubernet, M. Emprechtinger, P. Encrenaz, E. Falgarone, K. France, A. Fuente, M. Gerin, T. F. Giesen, A. M. di Giorgio, J. R. Goicoechea, P. F. Goldsmith, R. Güsten, A. Har-

ris, F. Helmich, E. Herbst, P. Hily-Blant, K. Jacobs, T. Jacq, C. Joblin, D. Johnstone, C. Kahane, M. Kama, T. Klein, A. Klotz, C. Kramer, W. Langer, B. Lefloch, C. Leinz, A. Lorenzani, S. D. Lord, S. Maret, P. G. Martin, J. Martin-Pintado, C. McCoey, M. Melchior, G. J. Melnick, K. M. Menten, B. Mookerjea, P. Morris, J. A. Murphy, D. A. Neufeld, B. Nisini, S. Pacheco, L. Pagani, B. Parise, J. C. Pearson, M. Pérault, T. G. Phillips, R. Plume, S.-L. Quin, R. Rizzo, M. Röllig, M. Salez, P. Saraceno, S. Schlemmer, R. Simon, K. Schuster, F. F. S. van der Tak, A. G. G. M. Tielens, D. Teyssier, N. Trappe, C. Vastel, S. Viti, V. Wakelam, A. Walters, S. Wang, N. Whyborn, M. van der Wiel, H. W. Yorke, S. Yu, and J. Zmuidzinas. Detection of interstellar oxidaniumyl: Abundant H_2O^+ towards the star-forming regions DR21, Sgr B2, and NGC6334. *Astronomy and Astrophysics*, 518:L111, 2010.

[36] B. Weis, S. Carter, P. Rosmus, H.-J. Werner, and P. J. Knowles. A theoretical rotationally resolved infrared spectrum for $H_2O^+(X^2B_1)$. *The Journal of Chemical Physics*, 91(5):2818–2833, 1989.

[37] S. E. Strahan, R. P. Mueller, and R. J. Saykally. Measurement of the rotational spectrum of the water cation (H_2O^+) by laser magnetic resonance. *The Journal of Chemical Physics*, 85(3):1252–1260, 1986.

[38] P. Mürtz, L. R. Zink, K. M. Evenson, and J. M. Brown. Measurement of high-frequency rotational transitions of H_2O^+ in its ground state by far-infrared laser magnetic resonance (LMR) spectroscopy. *The Journal of Chemical Physics*, 109(22):9744–9752, 1998.

[39] J. M. Brown and H. S. P. Müller. The rotational spectrum of the SH^+ radical in its $X^3\Sigma^-$ state. *Journal of Molecular Spectroscopy*, 255(1):68–71, 2009.

[40] S. Guilloteau, A. Baudry, C. M. Walmsley, T. L. Wilson, and A. Winnberg. Rotationally excited OH: Emission and absorption toward H II/OH regions. *Astronomy and Astrophysics*, 131:45–57, February 1984.

[41] M. Gerin, M. De Luca, J. Black, J. R. Goicoechea, E. Herbst, D. A. Neufeld, E. Falgarone, B. Godard, J. C. Pearson, D. C. Lis, T. G. Phillips, T. A. Bell, P. Sonnentrucker, F. Boulanger, J. Cernicharo, A. Coutens, E. Dartois, P. Encrenaz, T. Giesen, P. F. Goldsmith, H. C. Gupta, C. Gry, P. Hennebelle, P. Hily-Blant, C. Joblin, M. Kazmierczak, R. Kolos, J. Krelowski, J. Martin-Pintado, R. Monje, B. Mookerjea, M. Perault, C. Persson, R. Plume, P. B. Rimmer, M. Salez, M. Schmidt, J. Stutzki, D. Teyssier, C. Vastel, S. Yu, A. Contursi, K. Menten, T. Geballe, S. Schlemmer, R. Shipman, A. G. G. M. Tielens, S. Philipp-May, A. Cros, J. Zmuidzinas, L. A. Samoska, K. Klein, and A. Lorenzani. Interstellar OH^+, H_2O^+ and H_3O^+ along the sight-line to G10.6–0.4. *Astronomy and Astrophysics*, 518:L110, 2010.

[42] H. S. P. Müller, F. Schlöder, J. Stutzki, and G. Winnewisser. The Cologne Database for Molecular Spectroscopy, CDMS: a useful tool for astronomers and spectroscopists. *Journal of Molecular Structure*, 742(1–3):215–227, 2005.

[43] S. Yu, B. J. Drouin, J. C. Pearson, and H. M. Pickett. Terahertz spectroscopy and global analysis of H_3O^+. *The Astrophysical Journal Supplement Series*, 180(1):119–124, 2009.

[44] H. Gupta, P. Rimmer, J. C. Pearson, S. Yu, E. Herbst, N. Harada, E. A. Bergin, D. A. Neufeld, G. J. Melnick, R. Bachiller, W. Baechtold, T. A. Bell, G. A. Blake, E. Caux, C. Ceccarelli, J. Cernicharo, G. Chattopadhyay, C. Comito, S. Cabrit, N. R. Crockett, F. Daniel, E. Falgarone, M. C. Diez-Gonzalez, M.-L. Dubernet, N. Erickson, M. Emprechtinger, P. Encrenaz, M. Gerin, J. J. Gill, T. F. Giesen, J. R. Goicoechea, P. F. Goldsmith, C. Joblin, D. Johnstone, W. D. Langer, B. Larsson, W. B. Latter, R. H. Lin, D. C. Lis, R. Liseau, S. D. Lord, F. W. Maiwald, S. Maret, P. G. Martin, J. Martin-Pintado, K. M. Menten, P. Morris, H. S. P. Müller, J. A. Murphy, L. H. Nordh, M. Olberg, V. Ossenkopf, L. Pagani, M. Pérault, T. G. Phillips, R. Plume, S.-L. Qin, M. Salez, L. A. Samoska, P. Schilke, E. Schlecht, S. Schlemmer, R. Szczerba, J. Stutzki, N. Trappe, F. F. S. van der Tak, C. Vastel, S. Wang, H. W. Yorke, J. Zmuidzinas, A. Boogert, R. Güsten, P. Hartogh, N. Honingh, A. Karpov, J. Kooi, J.-M. Krieg, R. Schieder, and P. Zaal. Detection of OH^+ and H_2O^+ towards Orion KL. *Astronomy and Astrophysics*, 521:L47, 2010.

[45] H.-J. Werner, P. Rosmus, and E.-A. Reinsch. Molecular properties from MCSCF-SCEP wave functions. I. Accurate dipole moment functions of OH, OH^-, and OH^+. *The Journal of Chemical Physics*, 79(2):905–916, 1983.

[46] S. Wu, Y. Chen, X. Yang, Y. Guo, Y. Liu, Y. Li, R. J. Buenker, and P. Jensen. Vibronic transition moments and line intensities for H_2O^+. *Journal of Molecular Spectroscopy*, 225(1):96–106, 2004.

[47] P. Botschwina, P. Rosmus, and E.-A. Reinsch. Spectroscopic properties of the hydroxonium ion calculated from SCEP CEPA wavefunctions. *Chemical Physics Letters*, 102(4):299–306, 1983.

[48] M. de Val-Borro, P. Hartogh, J. Crovisier, D. Bockelée-Morvan, N. Biver, D. C. Lis, R. Moreno, C. Jarchow, M. Rengel, S. Szutowicz, M. Banaszkiewicz, F. Bensch, M. I. Blecka, M. Emprechtinger, T. Encrenaz, E. Jehin, M. Küppers, L.-M. Lara, E. Lellouch, B. M. Swinyard, B. Vandenbussche, E. A. Bergin, G. A. Blake, J. A. D. L. Blommaert, J. Cernicharo, L. Decin, P. Encrenaz, T. de Graauw, D. Hutsemékers, M. Kidger, J. Manfroid, A. S. Medvedev, D. A. Naylor, R. Schieder, D. Stam, N. Thomas, C. Waelkens, R. Szczerba, P. Saraceno, A. M. Di Giorgio, S. Philipp, T. Klein, V. Ossenkopf, P. Zaal, and R. Shipman. Water production in comet 81P/Wild 2 as determined by Herschel/HIFI. *Astronomy and Astrophysics*, 521:L50, 2010.

[49] J. Crovisier, P. Colom, E. Gérard, D. Bockelée-Morvan, and G. Bourgois. Observations at Nançay of the OH 18-cm lines in comets. *Astronomy and Astrophysics*, 393(3):1053–1064, 2002.

[50] T. L. Farnham and D. G. Schleicher. Physical and compositional studies of Comet 81P/Wild 2 at multiple apparitions. *Icarus*, 173(2):533–558,

2005.

[51] U. Fink, M. P. Hicks, and R. A. Fevig. Production rates for the stardust mission target: 81P/Wild 2. *Icarus*, 141(2):331–340, 1999.

[52] J. T. T. Mäkinen, J. Silén, W. Schmidt, E. Kyrölä, T. Summanen, J.-L. Bertaux, E. Quémerais, and R. Lallement. Water production of comets 2P/Encke and 81P/Wild 2 derived from SWAN observations during the 1997 apparition. *Icarus*, 152(2):268–274, 2001.

[53] Cernicharo, J., Waters, L. B. F. M., Decin, L., Encrenaz, P., Tielens, A. G. G. M., Agúndez, M., De Beck, E., Müller, H. S. P., Goicoechea, J. R., Barlow, M. J., Benz, A., Crimier, N., Daniel, F., Di Giorgio, A. M., Fich, M., Gaier, T., García-Lario, P., de Koter, A., Khouri, T., Liseau, R., Lombaert, R., Erickson, N., Pardo, J. R., Pearson, J. C., Shipman, R., Sánchez Contreras, C., and Teyssier, D. A high-resolution line survey of IRC + 10216 with Herschel/HIFI – First results: Detection of warm silicon dicarbide (SiC$_2$). *Astronomy and Astrophysics*, 521:L8, 2010.

CHAPTER 8

Appendix

CONTENTS

8.1　H$_2$O$^+$ SPECTROSCOPIC DATA FROM LEW

Reproduced with permission from Lew, *Canad. J. Phys.*, 54:2028, 1976 [1]. ©
1976 Canadian Science Publishing.

Table 8.1: Wavelengths and Frequencies of Transitions to $v'' = 0$: Σ subband

Branch	Rot. Trans.	L.Str.[†] (Int.)	(0,5,0) – (0,0,0)		(0,7,0) – (0,0,0)	
			λ_{air} (Å)	ν_{vac} cm^{-1}	λ_{air} (Å)	ν_{vac} cm^{-1}
$^P R_{1,N-1}$	$2_{02}-1_{10}$	0.50 (0.43)			6527.916 28.522	15314.594 313.173
	$3_{03}-2_{11}$	2.96 (2.18)	7446.836 46.836	13424.825 424.825	6524.922 25.242	15321.621 320.868
	$4_{04}-3_{12}$	1.46 (0.87)	– 7445.735	– 13426.809	6523.210 23.438	15325.641 325.106
	$5_{05}-4_{13}$	5.71 (2.66)	7446.603 46.603	13425.245 425.245	6524.407 24.146	15322.831 323.442
	$6_{06}-5_{14}$	2.28 (0.76)			6524.566 24.686	15322.457 322.174
$^P Q_{1,N}$	$1_{01}-1_{11}$	4.48 (4.26)	7466.552 67.763	13389.375 387.204	6541.477 41.765	15282.846 281.982
	$2_{02}-2_{12}$	2.47 (2.12)	7469.071 69.257	13384.859 384.526	6542.462 42.871	15280.628 279.589
	$3_{n3}-3_{13}$	10.28 (7.58)	7471.672 71.533	13380.201 380.449	6543.977 44.212	15277.007 276.453
	$4_{04}-4_{14}$	4.33 (2.59)	7474.003 74.003	13376.027 376.027	6545.025 45.151	15274.559 274.267
	$5_{05}-5_{15}$	15.50 (7.22)	7476.457 76.358	13371.638 371.815	6547.315 46.935	15269.218 270.105
	$6_{06}-6_{16}$	5.91 (2.01)			6547.165 47.165	15269.569 269.569
$^P P_{1,N-1}$	$0_{00}-1_{10}$	0.99 (0.99)			6550.693 51.084	15261.345 260.435[a]
	$1_{01}-2_{11}$	4.44 (4.22)	7494.344 95.457	13339.723 337.742	6562.802 63.105	15233.187* 232.484*
	$2_{02}-3_{12}$	1.95 (1.68)	7513.849 14.128	13305.096 304.600	6576.769 77.262	15200.837 199.697
	$3_{03}-4_{13}$	7.14 (5.26)	7535.352 35.352	13267.127 267.127	6592.792 93.125	15163.894 163.127
	$4_{04}-5_{14}$	2.73 (1.64)	7558.173 58.340	13227.070 226.778	6609.480 09.713	15125.607 125.074
	$5_{05}-6_{15}$	8.89 (4.14)	7582.229 82.229	13185.105 185.105	6628.238 28.019	15082.801 083.300
		3.03 (1.03)			6644.801 44.983	15045.205 044.794

Table 8.2: Σ Subband: Continued

Branch	Rot. Trans.	(0,9,0) – (0,0,0)		(0,11,0) – (0,0,0)	
		λ_{air} (Å)	ν_{vac} cm^{-1}	λ_{air} (Å)	ν_{vac} cm^{-1}
$^P R_{1,N-1}$	$2_{02}-1_{10}$	5787.120 87.488	17274.961 273.865	5183.853 84.420	19285.299 283.192
	$3_{03}-2_{11}$	5784.570 84.892	17282.576 281.614	5181.775 82.868	19293.034 288.966
	$4_{04}-3_{12}$	5783.335 83.611	17286.268 285.444	5180.859 –	19296.444 –
	$5_{05}-4_{13}$	5782.756 83.245	17287.997[b] 286.537[b]	5180.763 79.457	19296.801 301.666
	$6_{06}-5_{14}$	5782.756 83.245	17287.997[b] 286.537[b]		
$^P Q_{1,N}$	$1_{01}-1_{11}$	5797.676 97.963	17243.510 242.655	5192.438 92.130	19253.413 254.557
	$2_{02}-2_{12}$	5798.534 98.764	17240.957 240.273	5193.013 93.492	19251.284 249.508
	$3_{03}-3_{13}$	5799.559 99.781	17237.911 237.250	5193.792 94.827	19248.394 244.559
	$4_{04}-4_{14}$	5800.486 00.692	17235.157[b] 234.545	5194.628 96.736	19245.296 246.899
	$5_{05}-5_{15}$	5800.744 01.137	17234.391 233.222	5195.194 93.806	19243.202 248.343
	$6_{06}-6_{16}$	5800.486 00.906	17235.157[b] 233.908		
$^P P_{1,N-1}$	$0_{00}-1_{10}$	5804.917 05.220	17222.001 221.102[a]	5198.500 98.735	19230.962 230.092[a]
	$1_{01}-2_{11}$	5814.409 14.657	17193.886 193.152	5205.866 05.511	19203.752 205.061
	$2_{02}-3_{12}$	5825.484 25.764	17161.198 160.375	5214.625 15.137	19171.498 169.616
	$3_{03}-4_{13}$	5837.856 38.174	17124.831 123.895	5224.492 25.593	19135.354 131.257
	$4_{04}-5_{14}$	5851.057 51.348	17086.192 085.343	5235.145 37.370	19096.350 097.662
	$5_{05}-6_{15}$	5864.189 64.710	17047.932 46.417	5246.009 44.710	19056.806 061.527
		5876.984 77.576	17010.816 009.103		

Table 8.3: Σ Subband: Continued

Branch	Rot. Trans.	(0,13,0) – (0,0,0)		(0,15,0) – (0,0,0)	
		λ_{air} (Å)	ν_{vac} cm^{-1}	λ_{air} (Å)	ν_{vac} cm^{-1}
$^PR_{1,N-1}$	$2_{02}-1_{10}$	4684.621	21340.471	4268.015	23423.506
		84.805	339.635	-	-
	$3_{03}-2_{11}$	4682.380	21350.684	4266.218	23433.368
		82.498	350.147	66.380	432.481
	$4_{04}-3_{12}$	4680.432	21359.572		
		80.526	359.142		
	$5_{05}-4_{13}$	4678.199	21369.765	4263.261	23449.624
		78.703	367.465	63.497	448.323
	$6_{06}-5_{14}$				
$^PQ_{1,N}$	$1_{01}-1_{11}$	4691.612	21308.672	4273.892	23391.292
		91.815	307.748	-	-
	$2_{02}-2_{12}$	4692.100	21306.458		
		~	-		
	$3_{03}-3_{13}$	4692.197	21306.014	4274.360	23388.732
		-	-	74.479	388.085
	$4_{04}-4_{14}$				
	$5_{05}-5_{15}$	4689.963	21316.165		
		90.411	314.127		
	$6_{06}-6_{16}$				
$^PP_{1,N-1}$	$0_{00}-1_{10}$	4696.343	21287.207		
	$1_{01}-2_{11}$	4702.554	21259.090	4282.982	23341.650
		02.747	258.219	83.106	340.975
	$2_{02}-3_{12}$	4709.728	21226.708	4288.854	23309.694
		09.843	226.194	-	-
	$3_{03}-4_{13}$	4717.230	21192.951	4295.140	23275.580
		17.344	192.441	95.300	274.716
	$4_{04}-5_{14}$	4724.992	21159.561		
		24.794	159.026		
	$5_{05}-6_{15}$	4731.340	21129.749	4307.356	23209.569
		31.886	127.313	07.595	208.282

Note: [†]L. Str. – Calculated line strength including nuclear statistical weight; Int. – Calculated relative line intensities for $T = 500$ K and rotational energies of $v_2' = 7$ level. [*]Calculated from $^PQ_{1,N}(1)$ using known combination differences. Lines obscured. [b]Blended lines. [s]Satellite lines.

Table 8.4: Wavelengths and Frequencies of Transitions to $v'' = 0$: Δ Subband

Branch	Rot. Trans.	L.Str[†] (Int.)	(0,7,0) - (0,0,0) λ_{air} (Å)	(0,7,0) - (0,0,0) ν_{vac} cm^{-1}	(0,9,0) - (0,0,0) λ_{air} (Å)	(0,9,0) - (0,0,0) ν_{vac} cm^{-1}
$^r R_{1,N}$	$2_{21}-1_{11}$	4.48 (5.32)	6575.917 / 74.051	15202.806 / 207.120	5826.677 / 29.806	17157.685 / 148.476
	$3_{22}-2_{12}$	1.65 (1.66)	6568.750 / 67.298	15219.394 / 222.757	5821.577 / 23.520	17172.716 / 166.984
	$4_{23}-3_{13}$	5.37 (4.45)	6562.039 / 60.544	15234.957[b] / 238.429	5816.530 / 17.900	17187.615 / 183.568
	$5_{24}-4_{14}$	1.88 (1.20)	6554.914 / 53.462	15251.518 / 254.897	5811.476 / 12.287	17202.562 / 200.163
	$6_{25}-5_{15}$	5.75 (2.74)			5806.155 / 07.299	17218.329 / 214.936
$^r R_{1,N-1}$	$3_{21}-2_{11}$	4.94 (4.58)	6573.977 / 72.700	15207.292 / 210.247	5825.764 / 27.733	17160.375 / 154.575
	$4_{22}-3_{12}$	1.78 (1.47)			5825.121 / 26.494	17162.266 / 158.223
	$5_{23}-4_{13}$	5.47 (3.48)	6573.442 / 72.146	15208.531 / 211.527	5826.071 / 27.031	17159.471 / 156.643
$^r Q_{1,N}$	$3_{21}-3_{13}$	4.60 (4.69)	6593.328 / 91.942	15162.660 / 165.847	5840.977 / 42.860	17115.680 / 110.162
	$4_{22}-4_{14}$	2.28 (1.89)	6595.055 / 93.540	15158.689 / 162.171	5842.511 / 43.818	17111.185 / 107.357
	$5_{23}-5_{15}$	9.24 (5.89)	6596.710 / 95.286	15154.887 / 158.158	5844.346 / 45.217	17105.811 / 103.263
	$6_{24}-6_{16}$	3.93 (1.87)	6598.513 / 97.068	15150.746 / 154.064	5846.518 / 47.273	17099.459 / 097.250
	$7_{25}-7_{17}$	14.38 (4.75)			5849.162 / 49.520	17091.728 / 090.683
$^r Q_{1,N-1}$	$2_{21}-2_{11}$	2.47 (2.93)	6597.469 / 95.525	15153.143 / 157.608	5843.595 / 46.692	17108.010 / 098.949
	$3_{22}-3_{12}$	1.55 (1.58)	6603.377 / 01.962	15139.586 / 142.830	5848.765 / 50.757	17092.889 / 087.069
	$4_{23}-4_{13}$	7.06 (5.85)	6611.117 / 09.713	15121.861 / 125.074	5855.072 / 56.531	17074.476 / 070.223
	$5_{24}-5_{14}$	3.31 (2.11)	6619.711 / 18.223	15102.552 / 105.625	5862.242 / 63.163	17053.593 / 050.914
	$6_{25}-6_{15}$	13.31 (6.35)			5869.711 / 71.007	17031.893 / 028.133
$^p P_{3,N-2}$	$2_{21}-3_{31}$	7.35 (8.72)	6686.557 / 85.507	14951.252 / 953.600	5913.376 / 17.290	16906.129 / 894.946
	$3_{22}-4_{32}$	2.64 (2.69)	6700.178 / 6699.467	14920.857 / 922.441	5924.574 / 27.202	16874.174 / 866.692
	$4_{23}-5_{33}$	8.70 (7.20)	6715.495 / 14.650	14886.824 / 888.698	5936.791 / 38.772	16839.451 / 833.834
	$5_{24}-6_{34}$	3.19 (2.03)	6731.729 / 30.849	14850.925 / 852.866	5950.048 / 51.409	16801.933 / 798.090
	$6_{25}-7_{35}$	10.43 (4.98)			5964.060 / 65.750	16762.458 / 757.711
$^p P_{3,N-3}$	$3_{21}-4_{31}$	8.03 (8.18)	6700.524 / 6699.929	14920.087 / 921.412	5924.942 / 27.545	16873.127 / 865.718
	$4_{22}-5_{32}$	3.00 (2.48)	6717.036 / 16.193	14883.410 / 885.278	5938.060 / 39.973	16835.835 / 830.431
	$5_{23}-6_{33}$	10.13 (6.45)	6735.807 / 34.975	14841.935 / 843.767	5953.274 / 54.689	16792.828 / 788.837
	$6_{24}-7_{34}$	3.74 (1.78)	6757.563 / 56.658	14794.152 / 796.132	5971.061 / 72.311	16742.806 / 739.299
	$7_{25}-8_{35}$	12.06 (3.98)			5991.699 / 92.742	16685.134 / 682.232

Table 8.5: Δ Subband: Continued

Branch	Rot. Trans.	(0,11,0) – (0,0,0) λ_{air} (Å)	ν_{vac} cm^{-1}	(0,13,0) – (0,0,0) λ_{air} (Å)	ν_{vac} cm^{-1}	(0,15,0) – (0,0,0) λ_{air} (Å)	ν_{vac} cm^{-1}
$^rR_{1,N}$	$2_{21}-1_{11}$	5220.983 / 25.853	19148.152 / 130.305	4696.180 / 92.992	21287.947 / 302.406	4279.664 / 79.427	23359.749 / 361.043
	$3_{22}-2_{12}$	5217.689 / 22.569	19160.239 / 142.336				
	$4_{23}-3_{13}$	5213.300 / 18.955	19176.384 / 155.589	4687.733 / 85.243	21326.306 / 337.636		
	$5_{24}-4_{14}$						
	$6_{25}-5_{15}$						
$^rR_{1,N-1}$	$3_{21}-2_{11}$	5220.419 / 22.652	19150.219 / 142.030	4695.567 / 93.217	21290.723 / 301.385		
	$4_{22}-3_{12}$						
	$5_{23}-4_{13}$	5219.220 / 20.141	19154.619 / 151.238				
$^rQ_{1,N}$	$3_{21}-3_{13}$	5232.619 / 34.775	19105.571 / 097.703^b	4705.432 / 03.017	21246.091 / 257.000	4287.323	23318.019
	$4_{22}-4_{14}$						
	$5_{23}-5_{15}$	5233.875 / 34.721	19100.985 / 097.898	4708.104 / 04.930	21234.029 / 248.355	4290.030	23303.306
	$6_{24}-6_{16}$					–	–
	$7_{25}-7_{17}$						
$^rQ_{1,N-1}$	$2_{21}-2_{11}$	5234.564 / 39.415	19098.472 / 080.790	4707.148 / 03.915	21238.344 / 252.940		
	$3_{22}-3_{12}$						
	$4_{23}-4_{13}$	5244.231 / 50.024	19063.266 / 042.231	4712.722 / 10.272	21213.224 / 224.257		
	$5_{24}-5_{14}$						
	$6_{25}-6_{15}$						
$^pP_{3,N-2}$	$2_{21}-3_{31}$	5290.501 / 96.056	18896.542 / 876.723	4752.338 / 49.517	21036.392 / 048.884	4326.259 / 26.386	23108.159 / 107.484
	$3_{22}-4_{32}$	5300.278 / 05.794	18861.688 / 842.080				
	$4_{23}-5_{33}$	5309.712 / 16.033	18828.174 / 805.786	4765.534 / 63.331	20978.142 / 987.844		
	$5_{24}-6_{34}$						
	$6_{25}-7_{35}$						
$^pP_{3,N-3}$	$3_{21}-4_{31}$	5299.929 / 302.675	18862.929 / 853.161	4759.789 / 57.735	21003.463 / 012.530	4332.737	23073.614
	$4_{22}-5_{32}$						
	$5_{23}-6_{33}$	5321.073 / 22.356	18787.976 / 783.445	4778.529 / 75.599	20921.093 / 933.931	4348.430	22990.341
	$6_{24}-7_{34}$					–	–
	$7_{25}-8_{35}$						

Note: †L. Str. – Calculated line strength including nuclear statistical weight; Int. – Calculated relative line intensities for $T = 500\,\text{K}$ and rotational energies of $v'_2 = 8$ level. ^{b}Blended.

Table 8.6: Wavelengths and Frequencies of Transitions to $v'' = 0$: Π Subband

Branch	Rot. Trans.	L.Str.† (Int.)	(0,6,0) – (0,0,0)		(0,8,0) – (0,0,0)	
			λ_{air} (Å)	ν_{vac} cm⁻¹	λ_{air} (Å)	ν_{vac} cm⁻¹
$^rR_{0,N}$	$1_{10}-0_{00}$	3.00 (2.98)	6973.720 70.152	14335.596 342.934	6146.802 47.375	16264.120 262.605
	$2_{11}-1_{01}$	1.50 (1.34)	6965.470 63.331	14352.575 356.983	6140.528 40.917	16280.739 279.707
	$3_{12}-2_{02}$	5.59 (4.24)	6958.228 56.831	14367.513 370.397	6134.758 35.545	16296.053 293.961
	$4_{13}-3_{03}$	2.02 (1.23)	6951.234 50.339	14381.968 383.820	6129.025 29.356	16311.294 310.414
	$5_{14}-4_{04}$	5.82 (2.03)			6122.817 23.183	16327.834 326.855
	$6_{15}-5_{05}$	1.72 (0.58)			6115.694 16.116	16346.850 345.720
$^rQ_{0,N}$	$1_{11}-1_{01}$	1.50 (1.50)	6984.633 80.894	14313.198 320.864	6155.556 56.038	16240.991 239.720
	$2_{12}-2_{02}$	8.17 (7.43)	6987.737 85.267	14306.838 311.897	6158.637 58.862	16232.868 232.274
	$3_{13}-3_{03}$	4.21 (3.31)	6991.953 89.937	14298.213 302.337	6162.597 62.707	16222.436 222.146
	$4_{14}-4_{04}$	17.55 (11.41)	6996.352 94.687	14289.223 292.623	6166.921 66.961	16211.060 210.956
	$5_{15}-5_{05}$	7.47 (3.82)	6999.962 7000.394	14281.853 280.972	6171.139 71.139	16199.982 199.982
	$6_{16}-6_{06}$	26.83 (10.25)	7003.941 05.336	14273.740 269.539	6175.018 75.018	16189.801 189.801
	$7_{17}-7_{07}$	10.23 (2.77)			6178.408 78.297	16180.922 181.212
	$8_{18}-8_{08}$	34.01 (6.22)			6181.272 81.272	16173.425 173.425
$^rP_{0,N}$	$1_{10}-2_{02}$	1.05 (1.04)			6170.304 70.959	16202.173 200.455
	$2_{11}-3_{03}$	0.53 (0.47)			6178.944 79.395	16179.518 178.336
	$3_{12}-4_{04}$	1.63 (1.24)			6187.170 88.029	16158.008 155.764
	$4_{13}-5_{05}$	0.48 (0.29)			6194.596 94.970	16138.638 137.664
	$5_{14}-6_{06}$	1.16 (0.40)			6201.041 01.421	16121.865 120.876
$^PR_{2,N-1}$	$3_{13}-2_{21}$	0.16 (0.13)			6167.748 68.489	16208.887 206.940
	$4_{14}-3_{22}$	1.09 (0.71)			6166.305 -	16212.681 -
	$5_{15}-4_{23}$	0.57 (0.29)			6166.104 66.445	16213.208 212.314
	$6_{16}-5_{24}$	2.35 (0.90)			- 6167.043	- 16210.742
$^PR_{2,N-2}$	$3_{12}-2_{20}$	0.84 (0.64)			6162.872 64.329	16221.711 217.877
	$4_{13}-3_{21}$	0.80 (0.49)			6158.999 59.791	16231.914 229.827
	$5_{14}-4_{22}$	4.60 (1.61)			6156.726 57.444	16237.905 236.013
	$6_{15}-5_{23}$	2.38 (0.80)				

Table 8.7: Π Subband: Continued

Branch	Rot. Trans.	(0,10,0) – (0,0,0)		(0,12,0) – (0,0,0)		(0,14,0) – (0,0,0)	
		λ_{air} (Å)	ν_{vac} cm^{-1}	λ_{air} (Å)	ν_{vac} cm^{-1}	λ_{air} (Å)	ν_{vac} cm^{-1}
$^rR_{0,N}$	$1_{10}-0_{00}$	5478.548 81.217	18247.942 239.057	4922.382 17.854	20309.699 328.397	4469.358 68.823	22368.303 370.979
	$2_{11}-1_{01}$	5474.027 74.936	18263.012 259.978	4917.795 14.694	20328.643 341.470	4465.681 65.373	22386.717 388.261
	$3_{12}-2_{02}$	5469.107 70.140	18279.441 275.989	4913.294 10.829	20347.264 357.475	4462.123 61.952	22404.567 405.427
	$4_{13}-3_{03}$	5464.459 65.390	18294.989 291.874			4458.361 58.257	22423.474 423.998
	$5_{14}-4_{04}$	5459.530 60.129	18311.505[b] 309.498			4454.149 53.936	22444.680 445.754
	$6_{15}-5_{05}$	5453.486 53.971	18331.801 330.171				
$^rQ_{0,N}$	$1_{11}-1_{01}$	5486.156 87.794	18222.635 217.198	4927.756 23.016	20287.551 307.082	4474.350 73.815	22343.346 346.018
	$2_{12}-2_{02}$	5488.270 89.410	18215.618 211.834	4928.569 24.968	20284.203 299.034	4476.348 75.927	22333.375 335.474
	$3_{13}-3_{03}$	5491.333 91.917	18205.456 203.521	4928.766 25.204	20283.392 298.061	4478.698 78.288	22321.652 323.700
	$4_{14}-4_{04}$	5494.719 95.319	18194.238 192.252			4480.978 80.342	22310.297 313.462
	$5_{15}-5_{05}$	5498.208 98.666	18182.691 181.176			4482.904 82.278	22300.713 303.827
	$6_{16}-6_{06}$	5501.671 02.039	18171.247 170.034				
	$7_{17}-7_{07}$						
	$8_{18}-8_{08}$						
$^rP_{0,N}$	$1_{10}-2_{02}$	5497.220 99.952	18185.961 176.927	4937.433 32.837	20247.789 266.252	4481.762 81.256	22306.395 308.911
	$2_{11}-3_{03}$	5504.551 05.513	18161.741 158.566				
	$3_{12}-4_{04}$			4946.851 44.369	20209.241 219.384		
	$4_{13}-5_{05}$	5516.528 17.477	18122.311 119.193				
	$5_{14}-6_{06}$	5522.261	18103.497				
$^PR_{2,N-1}$	$3_{13}-2_{21}$	5496.492	18188.370				
	$4_{14}-3_{22}$	5494.239 95.214	18195.829 192.598				
	$5_{15}-4_{23}$	5494.239 94.950	18195.829 193.474				
	$6_{16}-5_{24}$	5495.151 95.747	18192.809 190.836				
$^PR_{2,N-2}$	$3_{12}-2_{20}$	5491.449 92.992	18205.074 199.958				
	$4_{13}-3_{21}$	5488.270 89.580	18215.618 211.270				
	$5_{14}-4_{22}$	5486.433 87.344	18221.715 218.690				
	$6_{15}-5_{23}$	5485.568 86.283	18224.590 222.213				

Table 8.8: Π Subband: Continued

Branch	Rot. Trans.	L.Str.[†] (Int.)	(0,6,0) – (0,0,0) λ_{air} (Å)	ν_{vac} cm^{-1}	(0,8,0) – (0,0,0) λ_{air} (Å)	ν_{vac} cm^{-1}
$^PQ_{2,N-1}$	$2_{11}-2_{21}$	0.82 (0.73)	7021.627 20.326	14237.787 240.426	6184.144 85.198	16165.914[b] 163.158
	$3_{12}-3_{22}$	4.25 (3.23)	7024.952 24.196	14231.049 232.580	6186.586 87.855	16159.533 156.218
	$4_{13}-4_{23}$	1.94 (1.18)	7029.134 28.687	14222.582 223.486	6189.527 90.246	16151.854 149.977
	$5_{14}-5_{24}$	7.22 (2.52)	7033.924 33.277	14212.896 214.204	6192.745 93.434	16143.463 141.666
	$6_{15}-6_{25}$	2.82 (0.94)			6195.950 96.658	16135.112 133.267
$^PQ_{2,N-2}$	$2_{12}-2_{20}$	1.73 (1.57)			6186.974 87.855	16158.520 156.218
	$3_{13}-3_{21}$	0.66 (0.52)			6192.890 -	16143.084 -
	$4_{14}-4_{22}$	1.44 (0.94)			6201.280 01.719	16121.243 120.101
	$5_{15}-5_{23}$	0.24 (0.12)			6212.536 -	16092.033 -
$^PP_{2,N-1}$	$1_{11}-2_{21}$	1.47 (1.47)	7041.117 38.188	14198.378 204.286	6199.388 200.543	16126.163 123.158
	$2_{12}-3_{22}$	4.86 (4.42)	7055.051 53.196	14170.335 174.062	6210.869 11.589	16096.352 094.487
	$3_{13}-4_{23}$	1.80 (1.42)	7070.807 -	14138.758 -	6223.756 24.271	16063.023 061.696
	$4_{14}-5_{24}$	5.91 (3.84)	7087.822 86.519	14104.819 107.411	6237.838 38.212	16026.762 025.800
	$5_{15}-6_{25}$	2.13 (1.09)	7105.311 06.109	14070.101 068.520	6252.846 53.127	15988.294 987.577
	$6_{16}-7_{26}$	6.75 (2.58)	7124.388 26.824	14032.424 027.629	6268.459 68.660	15948.474 947.962
	$7_{17}-8_{27}$	2.34 (0.63)			6284.246 84.378	15908.409 908.074
	$8_{18}-9_{28}$	7.17 (1.31)			6299.990 300.236	15868.652 868.033
$^PP_{2,N-2}$	$1_{10}-2_{20}$	4.85 (4.82)	7040.646 37.883	14199.326 204.902	6198.747 200.030	16127.829 124.493
	$2_{11}-3_{21}$	2.07 (1.85)	7054.249 52.723	14171.946 175.012	6209.417 10.328	16100.117 097.756
	$3_{12}-4_{22}$	8.08 (6.13)	7070.407 69.496	14139.558 141.380	6221.774 22.275	16068.142 066.847
	$4_{13}-5_{23}$	3.38 (2.06)	7089.142 88.660	14102.191 103.150	6235.995 36.693	16031.498 029.704
	$5_{14}-6_{24}$	12.07 (4.21)	7110.321 09.681	14060.186 061.452	6251.862 52.571	15990.812 988.997
	$6_{15}-7_{25}$	4.57 (1.53)			6268.982 69.236	15947.142 945.225

Table 8.9: Π Subband: Continued

Branch	Rot. Trans.	(0,10,0) – (0,0,0)		(0,12,0) – (0,0,0)		(0,14,0) – (0,0,0)	
		λ_{air} (Å)	ν_{vac} cm^{-1}	λ_{air} (Å)	ν_{vac} cm^{-1}	λ_{air} (Å)	ν_{vac} cm^{-1}
$^PQ_{2,N-1}$	$2_{11}-2_{21}$	5508.680	18148.128[b]				
		10.129	143.356				
	$3_{12}-3_{22}$	5510.273	18142.882	4946.465	20210.816		
		11.713	138.140	44.264	219.814		
	$4_{13}-4_{23}$	5512.506	18135.531				
		13.757	131.418				
	$5_{14}-5_{24}$	5515.047	18127.175				
		15.921	124.304				
	$6_{15}-6_{25}$	5517.237	18119.982				
		17.927	117.716				
$^PQ_{2,N-2}$	$2_{12}-2_{20}$	5510.758	18141.284	4946.682	20209.931	4491.292	22259.063
		12.443	135.783	43.482	223.013	91.221	259.414
	$3_{13}-3_{21}$	5515.384	18126.070			4494.696	22242.204
		16.354	122.883			94.514	243.105
	$4_{14}-4_{22}$	5521.999	18104.355			4499.099	22220.442
		22.904	101.387			98.653	222.642
	$5_{15}-5_{23}$						
$^PP_{2,N-1}$	$1_{11}-2_{21}$	5520.951	18107.791	4955.799	20172.753	4497.461	22228.532
		23.129	100.653	51.426	190.568[b]	97.286	229.395
	$2_{12}-3_{22}$	5529.717	18079.086	4961.963	20147.691	4503.878	22196.860
		31.268	074.016	58.638	161.204	03.720	197.643
	$3_{13}-4_{23}$	5539.848	18046.025	4967.817	20123.950	4510.933	22162.148
		40.758	043.062	64.425	137.699[b]	10.846	163.192
	$4_{14}-5_{24}$	5550.957	18009.910			4518.320	22125.913
		51.832	007.072			17.842	128.254
	$5_{15}-6_{25}$	5562.979	17970.988			4525.864	22089.035
		63.673	968.749			25.364	091.474
	$6_{16}-7_{26}$	5575.701	17929.986				
		76.311	928.025				
	$7_{17}-8_{27}$						
	$8_{18}-9_{28}$						
$^PP_{2,N-2}$	$1_{10}-2_{20}$	5519.786	18111.613	4955.638	20173.406	4496.757	22232.014
		23.057	100.887	51.494	190.292	96.592	232.829
	$2_{11}-3_{21}$	5528.724	18082.332	4961.890	20147.989	4502.000	22206.121
		30.052	077.990	59.064	159.471	01.973	206.257
	$3_{12}-4_{22}$	5538.184	18051.446	4968.973	20119.271	4508.001	22176.559
		39.562	046.955	66.711	128.430	08.001[b]	176.559[b]
	$4_{13}-5_{23}$	5549.328	18015.196			4514.714	22143.587
		50.568	011.173			14.758	143.369
	$5_{14}-6_{24}$	5561.844	17974.657[b]			4522.040	22107.714
		62.773	971.656			22.001	107.903
	$6_{15}-7_{25}$	5575.064	17932.036				
		75.798	929.674				

Table 8.10: Φ Subband

Branch	Rot. Trans.	L.Str.[†] (Int.)	(0,8,0) – (0,0,0) λ_{air} (Å)	ν_{vac} cm^{-1}	(0,10,0) – (0,0,0) λ_{air} (Å)	ν_{vac} cm^{-1}
$^rR_{2,N-1}$	$3_{31}-2_{21}$	2.45 (3.62)	6251.862 53.598	15990.812 986.373		
	$4_{32}-3_{22}$	7.65 (9.19)	6248.042 49.150	16000.587 997.751	5561.844 66.107	17974.657 960.891
	$5_{33}-4_{23}$	2.64 (2.44)	6244.940 45.607	16008.537 006.826	5560.038 62.979	17980.494 970.988
	$6_{34}-5_{24}$	8.01 (5.39)	6242.173 42.557	16015.632 014.647	5558.887 60.633	17984.218 978.573
$^rR_{2,N-2}$	$3_{30}-2_{20}$	7.29 (10.78)	6252.030 53.792	15990.383 985.878	5564.493 70.711	17966.100 946.047
	$4_{31}-3_{21}$	2.45 (2.94)	6249.617 50.732	15996.557 993.702	5564.024[b] 68.437	17967.614[b] 953.377
	$5_{32}-4_{22}$	7.16 (6.61)	6248.667 49.343	15998.987 997.257	5562.773 65.895	17971.656 961.575
	$6_{33}-5_{23}$	2.21 (1.49)	6249.904 50.257	15995.821 994.918	5561.844 63.673	17974.657 968.749
$^rQ_{2,N-1}$	$3_{30}-3_{22}$	2.55 (3.77)	6276.436 78.053	15928.204 924.100	5583.826 89.969	17903.895 884.221
	$4_{31}-4_{23}$	1.62 (1.95)	6281.056 82.100	15916.487 913.844	5588.924 93.326	17887.566 873.486
	$5_{32}-5_{24}$	7.21 (6.65)	6285.792 86.426	15904.495 902.891	5592.149 95.292	17877.249 867.206
	$6_{33}-6_{25}$	3.24 (2.18)	6291.033 91.336	15891.245 890.480	5594.390 96.186	17870.090 864.352
$^rQ_{2,N-2}$	$3_{31}-3_{21}$	0.82 (1.21)	6277.671 79.280	15925.070 920.988		
	$4_{32}-4_{22}$	4.41 (5.30)	6283.953 84.991	15909.151 906.522	5590.282 94.530	17883.220 869.640
	$5_{33}-5_{23}$	2.02 (1.86)	6292.257 92.893	15888.154 886.549	5597.527 –	17860.073 –
	$6_{34}-6_{24}$	7.61 (5.12)	6302.271 02.652	15862.909 861.951	5606.503 08.253	17831.481 825.915
$^pP_{4,N-3}$	$3_{31}-4_{41}$	3.30 (4.88)	6395.762 98.356	15631.033[b] 624.697[b]		
	$4_{32}-5_{42}$	10.33 (12.41)	6409.657 11.521	15597.147 592.614	5689.566 94.584	17571.159 555.675
	$5_{33}-6_{43}$	3.65 (3.37)	6424.690 26.036	15560.654 557.392	5702.082 05.672	17532.590 521.560
	$6_{34}-7_{44}$	11.67 (7.85)	6440.730 41.722	15521.902 519.510	5715.820 18.124	17490.451 483.404
$^pP_{4,N-4}$	$3_{30}-4_{40}$	9.91 (14.66)	6395.762 98.356	15631.033[b] 624.697[b]	5678.065 85.146	17606.749 584.818
	$4_{31}-5_{41}$	3.45 (4.14)	6410.096 11.975	15596.080 591.526	5690.870 96.062	17567.131 551.120
	$5_{32}-6_{42}$	10.97 (10.13)	6425.267 26.626	15559.256 555.964	5702.275 06.094	17531.995 520.262
	$6_{33}-7_{43}$	3.91 (2.63)	6441.998 42.930	15518.845 516.600	5713.445 15.820	17497.721 490.451
	$7_{34}-8_{44}$	12.45 (5.83)	6460.258 60.828	15474.982 473.615		

8.2 D_2O^+ SPECTROSCOPIC DATA FROM LEW AND GROLEAU

Reproduced with permission from H. Lew and R. Groleau, *Canad. J. Phys.*, 65:739, 1987 [2]. © 1987 Canadian Science Publishing.

Table 8.11: Wavelengths and Frequencies of Transitions to $v'' = 0$: Σ Subband

Branch	Rot. Trans.	L.Str. (Int.)*	(0,11,0) − (0,0,0) λ_{air} (Å)	ν_{vac} (cm⁻¹)	(0,13,0) − (0,0,0) λ_{air} (Å)	ν_{vac} (cm⁻¹)
$^PR_{1,N-1}$	$2_{02}-1_{10}$	2.99 (2.41)	6292.546 / 93.137	15887.424 / 885.933	5766.457 / 66.662	17336.861 / 336.247
	$3_{03}-2_{11}$	2.98 (1.93)	6290.915	15891.543	5765.042 / 65.282	17341.118 / 340.395
	$4_{04}-3_{12}$	8.88 (4.29)	6289.999 / 90.149	15893.858 / 893.479	5764.369 / 64.559	17343.143 / 342.570
	$5_{05}-4_{13}$	5.84 (1.94)			5764.264 / 64.432	17343.459 / 342.952
	$6_{06}-5_{14}$	14.24 (3.00)			5764.264 / 64.369	17343.459 / 343.143
$^PQ_{1,N}$	$1_{01}-1_{11}$	4.49 (4.18)	6298.994 / 99.180	15871.162 / 870.693	5771.870 / 72.022	17320.603 / 320.147
	$2_{02}-2_{12}$	14.91 (12.02)	6299.559 / 300.068	15869.739 / 868.456	5772.352 / 72.485	17319.158 / 318.759
	$3_{03}-3_{13}$	10.38 (6.74)	6300.279 / 300.279	15867.925 / 867.925	5772.916 / 73.112	17317.466 / 316.878
	$4_{04}-4_{14}$	26.43 (12.77)	6300.952 / 01.047	15866.229 / 865.991	5773.560 / 73.731	17315.534 / 315.021
	$5_{05}-5_{15}$	15.91 (5.28)			5774.126 / 74.239	17313.838 / 313.499
	$6_{06}-6_{16}$	38.86 (7.78)			5774.330 / 74.401	17313.226 / 313.011
$^PP_{1,N-1}$	$0_{00}-1_{10}$	5.98 (5.98)	6303.310 / 03.499	15860.294[s] / 859.820[s]	5775.467 / 75.629	17309.817[s] / 309.331[s]
	$1_{01}-2_{11}$	4.47 (4.16)	6308.904 / 09.054	15846.232 / 845.856	5780.200 / 80.332	17295.642 / 295.250
	$2_{02}-3_{12}$	11.84 (9.54)	6315.445 / 15.970	15829.820 / 828.503	5785.704 / 85.853	17279.189 / 278.744
	$3_{03}-4_{13}$	7.30 (4.74)	6322.785 / 22.833	15811.443 / 811.323	5791.823 / 92.056	17260.933 / 260.241
	$4_{04}-5_{14}$	17.09 (8.25)	6330.641 / 30.791	15791.822 / 792.448	5798.503 / 98.721	17241.050 / 240.401
	$5_{05}-6_{15}$	9.54 (3.17)			5805.605	17219.958
	$6_{06}-7_{16}$	20.32 (4.28)			5812.278 / 12.423	17200.189 / 199.759

Table 8.12: Wavelengths and Frequencies of Transitions to $v'' = 0$: Σ Subband, Continued

Branch	Rot. Trans.	$(0,15,0) - (0,0,0)$		$(0,17,0) - (0,0,0)$	
		λ_{air} (Å)	ν_{vac} (cm⁻¹)	λ_{air} (Å)	ν_{vac} (cm⁻¹)
$^PR_{1,N-1}$	$2_{02}-1_{10}$	5314.209 14.403	18812.241 811.555	4923.088	20306.787
	$3_{03}-2_{11}$	5313.079 13.210	18816.244 815.781	4922.094	20310.888
	$4_{04}-3_{12}$	5312.404 12.582	18818.635 818.003	4921.561	20313.087
	$5_{05}-4_{13}$	5312.103 12.311	18819.702 818.964	4921.379	20313.838
	$6_{06}-5_{14}$	5311.965 12.103	18820.189 819.702		
$^PQ_{1,N}$	$1_{01}-1_{11}$	5318.811 18.953	18795.966 795.462	4927.158 27.401	20290.011 289.012
	$2_{02}-2_{12}$	5319.197 19.338	18794.601 794.104	4927.401 27.270	20289.012 289.551
	$3_{03}-3_{13}$	5319.844	18792.315	4927.823	20287.273
	$4_{04}-4_{14}$	5320.120 20.340	18791.059 790.563	4928.250	20285.515
	$5_{05}-5_{15}$	5320.499 20.674	18790.002 789.384	4928.592	20284.110
	$6_{06}-6_{16}$	5320.499	18790.002		
$^PP_{1,N-1}$	$0_{00}-1_{10}$	5321.860 22.001	18785.196 784.700ˢ		
	$1_{01}-2_{11}$	5325.873 25.992	18771.043 770.625	4933.217 33.444	20265.091 264.159
	$2_{02}-3_{12}$	5330.534 30.712	18754.630 754.004	4937.109 36.998	20249.119 249.576
	$3_{03}-4_{13}$	5335.809 35.923	18736.088 735.690	4941.590	20230.756
	$4_{04}-5_{14}$	5341.364 41.556	18716.604 715.932	4946.386	20211.140
	$5_{05}-6_{15}$	5347.103 47.326	18696.515 695.736		
	$6_{06}-7_{16}$	5352.716 52.879	18676.910 676.340		

Note: ˢStands for shoulder; Rot. Trans., rotational transition; L. Str. (Int.), line strength (intensity). *Intensities calculated for $T = 180$ K and rotational levels of $\tilde{A}$ (0,15,0) relative to 0_{00}.

Table 8.13: Wavelengths and Frequencies of Transitions to $v'' = 1$: Σ Subband

Branch	Rot. Trans.	(0,13,0) - (0,1,0)		(0,15,0) - (0,1,0)		(0,17,0) - (0,1,0)	
		λ_{air} (Å)	ν_{vac} (cm⁻¹)	λ_{air} (Å)	ν_{vac} (cm⁻¹)	λ_{air} (Å)	ν_{vac} (cm⁻¹)
$^P R_{1,N-1}$	$2_{02}-1_{10}$	6136.667	16290.983	5627.035	17766.418	5190.444	19260.810
		36.949	290.234	27.304	765.532	90.520	260.527
	$3_{03}-2_{11}$	6135.087	16295.178			5189.334	19264.934
		35.380	294.399				
	$4_{04}-3_{12}$	6134.323	16297.208	5625.050	17772.685	5188.821	19266.835
		34.551	296.852	25.280	771.961		
	$5_{05}-4_{13}$	6134.323	16297.208				
		34.457	296.852				
	$6_{06}-5_{14}$	6134.323	16297.208	5624.610	17774.078		
		34.457	296.852	24.803	773.467		
$^P Q_{1,N}$	$1_{01}-1_{11}$	6142.753	16274.842	5632.203	17750.117	5194.921	19244.212
		43.043	274.074	32.404	749.482	95.079	243.627
	$2_{02}-2_{12}$	6143.260	16273.500	5632.596	17748.877	5195.146	19243.380
		43.430	273.047	32.774	748.315	95.146	243.380
	$3_{03}-3_{13}$	6143.816	16272.025			5195.579	19241.775
		44.053	271.399			95.683	241.389
	$4_{04}-4_{14}$	6144.455	16270.333	5633.573	17745.798	5195.971	19240.323
		44.663	269.789	33.752	745.236		
	$5_{05}-5_{15}$	6145.074ˢ	16268.696ˢ				
		45.154ˢ	267.484ˢ				
	$6_{06}-6_{16}$	6145.074ˢ	16268.696ˢ				
		45.154ˢ	267.484ˢ				
$^P P_{1,N-1}$	$0_{00}-1_{10}$	6146.862	16263.963				
	$1_{01}-2_{11}$	6152.227	16249.781	5640.160	17725.075	5201.689	19219.172
		52.408	249.302	40.311	724.599	01.921	218.318
	$2_{02}-3_{12}$	6158.482	16233.273	5645.392	17708.646ˢ	5206.030	19203.149
		58.709	232.676	45.564	708.106ˢ	06.030	203.149
	$3_{03}-4_{13}$	6165.464	16214.894			5211.018	19184.766
		65.728	214.197			11.172	184.199
	$4_{04}-5_{14}$	6173.070	16194.915	5657.595	17670.450	5216.404	19164.959
		73.339	194.208	57.825	669.732		
	$5_{05}-6_{15}$	6181.186	16173.650	5664.093	17650.178		
	$6_{06}-7_{16}$	6188.877	16153.551	5670.459	17630.364		
		89.052	153.095	70.557	630.061		

Table 8.14: Wavelengths and Frequencies of Transitions to $v'' = 0$ and 1: Δ Subband

Branch	Rot. Trans.	L. Str. (Int.)*	(0,13,0) – (0,0,0) λ_{air} (A)	ν_{vac} (cm^{-1})	(0,15,0) – (0,0,0) λ_{air} (A)	ν_{vac} (cm^{-1})	(0,15,0) – (0,1,0) λ_{air} (A)	ν_{vac} (cm^{-1})
$^rR_{1,N}$	$2_{21}-1_{11}$	4.49 (5.81)	5782.736 82.201	17288.059 289.657	5329.502 31.589	18758.263 750.918	5646.558 44.799	17704.992 710.508
	$3_{22}-2_{12}$	9.94 (10.29)	5780.092 79.696	17295.967 297.151	5327.472 28.785	18765.409 760.786	5641.901 43.376	17719.605 714.974
	$4_{23}-3_{13}$	5.45 (4.19)			5325.278 26.380	18773.142 769.257	5639.095 40.595	17728.421 723.707
	$5_{24}-4_{14}$	11.63 (6.25)	5775.139	17310.800	5323.725 24.283	18778.617 776.648	5637.548 38.177	17733.288 731.309
$^rR_{1,N-1}$	$2_{20}-1_{10}$	8.97 (11.37)	5783.373 82.849	17286.156 287.722	5330.025 32.127	18756.421 749.026	5644.799 47.197	17710.508 702.986
	$3_{21}-2_{11}$	4.97 (5.15)	5782.045 81.656	17290.124 291.289	5329.086 30.387	18759.726 755.146	5643.773 45.262	17713.727 709.054
	$4_{22}-3_{12}$	10.85 (8.44)	5781.656 81.317	17291.289 292.303	5328.909 29.800	18760.350 757.213	5643.575 44.596	17714.349 711.145
$^rQ_{1,N}$	$2_{20}-2_{12}$	4.97 (6.43)	5789.306 88.706	17268.439 270.228	5335.063 37.096	18738.708 731.572	5650.366 52.712	17693.057 685.716
	$3_{21}-3_{13}$	4.59 (4.76)	5789.963 89.529	17266.479 267.775	5335.809 37.096	18736.088 731.572	5651.139 52.602	17690.637 686.061
	$4_{22}-4_{14}$	13.55 (10.54)	5790.907 90.533	17263.664 264.781	5336.771 37.631	18732.713 729.695	5652.147 53.131	17687.484 684.406
$^rQ_{1,N-1}$	$2_{21}-2_{11}$	2.48 (3.21)	5791.098 90.533	17263.096 264.781	5336.605 38.686	18733.295 725.992	5654.539 52.712	17680.001 685.716
	$3_{22}-3_{12}$	9.23 (9.55)	5793.479 93.100	17256.000 257.130	5338.839 40.173	18725.456 720.779	5654.680 56.217	17679.559 674.757
	$4_{23}-4_{13}$	6.94 (5.34)			5341.364 42.492	18716.604 712.543	5657.360 58.833	17671.184 666.585
	$5_{24}-5_{14}$	19.31 (10.37)	5800.091	17236.329	5344.938 45.535	18704.091 702.001	5661.594 62.282	17657.970 655.826
$^pP_{3,N-2}$	$2_{21}-3_{31}$	7.47 (9.67)	5829.538 29.366	17149.264 149.771	5369.247 71.681	18619.409 610.971		
	$3_{22}-4_{32}$	16.10 (16.66)	5834.856 34.789	17133.633 133.830	5373.984 75.603	18602.994 597.391	5698.082 5700.023	17544.897 538.924
	$4_{23}-5_{33}$	8.87 (6.82)			5378.944 80.296	18585.840 581.170	5703.603 05.235	17527.916 522.900
	$5_{24}-6_{34}$	19.64 (10.55)	5847.191	17097.490	5384.918 85.720	18565.223 562.457		
$^pP_{3,N-3}$	$2_{20}-3_{30}$	14.96 (19.36)	5829.538 29.366	17149.264 149.771	5369.247 71.681	18619.409 610.971	5692.779 95.666	17561.242 552.339
	$3_{21}-4_{31}$	8.13 (8.42)	5835.004 34.948	17133.083 133.364	5374.098 75.683	18602.602 597.117	5698.201 5700.105	17544.530 538.672
	$4_{22}-5_{32}$	18.20 (14.16)	5841.377 41.297	17114.506 114.743	5379.635 80.763	18583.453 579.560	5704.371 05.738	17525.556 521.354
$^rP_{1,N}$	$3_{22}-4_{14}$	1.40 (1.45)	5802.762	17228.395				
$^rP_{1,N-1}$	$2_{20}-3_{12}$	0.68 (0.88)	5802.705	17228.564				

*Intensities calculated for $T = 180$ K and rotational levels of $\tilde{A}$ (0,15,0) relative to 0_{00}.

Table 8.15: Wavelengths and Frequencies of Transitions to $v'' = 0$: Π Subband

Branch	Rot. Trans.	L. Str. (Int.)*	(0,10,0) – (0,0,0) λ_{air} (A)	ν_{vac}(cm^{-1})	(0,12,0) – (0,0,0) λ_{air} (A)	ν_{vac}(cm^{-1})
$^{r}R_{0,N}$	1_{10}–0_{00}	3.00 (2.97)			6022.205	16600.618
	2_{11}–1_{01}	8.99 (7.63)			6018.974 18.152	16609.528 611.796
	3_{12}–2_{02}	5.69 (3.84)			6016.217 15.622	16617.137 618.782
	4_{13}–3_{03}	12.73 (6.30)			6013.527 13.183	16624.573 625.522
	5_{14}–4_{04}	6.41 (2.15)			6010.765 10.370	16632.210 633.304
	6_{15}–5_{05}	11.91 (2.51)			6007.598 07.469	16640.978 641.335
$^{r}Q_{0,N}$	1_{11}–1_{10}	8.99 (8.99)	6598.653 98.292	15150.425 151.252	6026.778 25.788	16588.020 590.746
	2_{12}–2_{02}	8.07 (7.05)	6599.720 99.663	15147.974 148.104	6028.272 27.476	16583.909 586.099
	3_{13}–3_{03}	24.73 (17.66)	6601.233 01.417	15144.502 144.080	6029.589 28.889	16580.287 582.213
	4_{14}–4_{04}	17.27 (9.41)			6031.892 31.178	16573.957 575.919
	5_{15}–5_{05}	44.74 (17.36)			6033.950 33.350	16568.303 569.952
	6_{16}–6_{06}	27.26 (7.00)				
	7_{17}–7_{07}	63.48 (10.09)			6038.356 38.356	16556.214^{b} 556.214^{b}
$^{P}R_{2,N-2}$	4_{13}–3_{21}	4.33 (2.14)			6029.762	16579.810
	6_{15}–5_{23}	13.37 (2.82)			6027.874 27.874	16585.005^{b} 585.005^{b}

^{b}stands for blended; ^{s}for shoulder. *Intensities calculated for $T = 180$ K and rotational levels of $\tilde{A}$ (0,14,0) relative to 1_{11}.

Table 8.16: Wavelengths and Frequencies of Transitions to $v'' = 0$: Π Subband Continued

Branch	Rot. Trans.	L. Str. (Int.)*	λ_{air} (A) (0,10,0)−(0,0,0)	ν_{vac} (cm⁻¹) (0,10,0)−(0,0,0)	λ_{air} (A) (0,12,0)−(0,0,0)	ν_{vac} (cm⁻¹) (0,12,0)−(0,0,0)
$^PQ_{2,N-1}$	$2_{11}-2_{21}$	4.95 (4.20)			6042.123 41.633	16545.892 547.236
	$4_{13}-4_{23}$	11.88 (5.88)			6044.805 44.651	16538.552[b] 538.974[b]
	$6_{15}-6_{25}$	17.64 (3.72)			6048.194 48.194	16529.284[b] 529.284[b]
$^PQ_{2,N-2}$	$3_{13}-3_{21}$	4.72 (3.37)	6620.816 21.288	15099.709 098.632	6045.918 45.456	16535.508 536.771
	$5_{15}-5_{23}$	2.53 (0.98)				
$^PP_{2,N-1}$	$1_{11}-2_{21}$	8.90 (8.90)	6626.487 26.552	15086.786 086.637	6049.994 49.349	16524.366 526.129
	$2_{12}-3_{22}$	4.92 (4.30)	6632.400 32.647	15073.337 072.774	6055.540 54.974	16509.233 510.775
	$3_{13}-4_{23}$	11.00 (7.85)	6638.965 39.386	15058.430 057.476	6061.046 60.536	16494.235 495.624
	$4_{14}-5_{24}$	6.09 (3.32)			6067.892 67.331	16475.626 477.150
	$5_{15}-6_{25}$	13.30 (5.16)			6074.916 74.443	16456.576 457.859
	$6_{16}-7_{26}$	7.16 (1.84)				
	$7_{17}-8_{27}$	15.17 (2.41)				
$^PP_{2,N-2}$	$1_{10}-2_{20}$	4.81 (4.77)			6049.349 48.483	16526.129 528.495
	$2_{11}-3_{21}$	12.15 (10.32)			6054.240 53.672	16512.778 514.328
	$3_{12}-4_{22}$	7.88 (5.31)			6060.064 59.681	16496.908 497.951
	$4_{13}-5_{23}$	19.99 (9.90)			6066.681 66.509	16478.915 479.383
	$5_{14}-6_{24}$	12.17 (4.09)			6074.149 73.922	16458.656 459.270
	$6_{15}-7_{25}$	28.42 (6.00)			6082.256 92.256	16436.717[b] 436.717

Table 8.17: Wavelengths and Frequencies of Transitions to $v'' = 0$: Π Subband Continued

Branch	Rot. Trans.	$(0,14,0) - (0,0,0)$		$(0,16,0) - (0,0,0)$	
		λ_{air} (Å)	ν_{vac}(cm^{-1})	λ_{air} (Å)	ν_{vac}(cm^{-1})
$^rR_{0,N}$	$1_{10}-0_{00}$	5534.579 35.049	18063.204 061.670[s]	5114.500 15.483	19546.806 543.049[s]
	$2_{11}-1_{01}$	5531.995 32.313	18071.642 070.604	5112.178 12.717	19555.684 553.622
	$3_{12}-2_{02}$	5529.627 29.914	18079.379 078.444	5110.184 10.626	19563.317 561.623
	$4_{13}-3_{03}$	5527.319 27.537	18086.930 086.218	5108.418 08.691	19570.080 569.031
	$5_{14}-4_{04}$	5524.912 25.116	18094.811 094.142		
	$6_{15}-5_{05}$	5522.087 22.361	18104.066 103.169		
$^rQ_{0,N}$	$1_{11}-1_{10}$	5538.242 38.649	18051.258 049.932		
	$2_{12}-2_{02}$	5539.605 39.793	18046.815 046.205		
	$3_{13}-3_{03}$	5541.386 41.519	18041.017 040.584	5120.996	19522.012
	$4_{14}-4_{04}$	5543.446 43.532	18034.311 034.031		
	$5_{15}-5_{05}$	5545.598 45.621	18027.315 027.239	5124.071 24.166	19510.298 509.935
	$6_{16}-6_{06}$	5547.674 47.674	18020.568[b] 020.568[b]	5125.350 25.798	19505.427 503.722
	$7_{17}-7_{07}$	5549.566 49.566	18014.425[b] 014.425[b]	5127.509 27.670	19497.213 496.601
$^pR_{2,N-2}$	$4_{13}-3_{21}$	5541.028	18042.182	5120.564	19523.660
	$6_{15}-5_{23}$	5539.221 39.605	18048.068 046.815		

Table 8.18: Wavelengths and Frequencies of Transitions to $v'' = 0$: Π Subband
Continued

Branch	Rot. Trans.	(0,14,0) − (0,0,0)		(0,16,0) − (0,0,0)	
		λ_{air} (Å)	ν_{vac} (cm^{-1})	λ_{air} (Å)	ν_{vac} (cm^{-1})
$^PQ_{2,N-1}$	$2_{11}-2_{21}$	5551.540	18008.019	5128.877	19492.013
		52.152	006.034	29.645	489.096
	$4_{13}-4_{23}$	5553.732	18000.910	5130.979	19484.028
		54.124	17999.642	31.387	482.481
	$6_{15}-6_{25}$	5556.368	17992.372		
		56.764	991.091		
$^PQ_{2,N-2}$	$3_{13}-3_{21}$	5555.174	17996.372		
		55.508	995.156		
	$5_{15}-5_{23}$	5562.827	17971.482		
$^PP_{2,N-1}$	$1_{11}-2_{21}$	5557.829	17987.642		
		58.534	985.360		
	$2_{12}-3_{22}$	5562.589	17972.249		
		63.005	970.905		
	$3_{13}-4_{23}$	5567.934	17954.998[b]	5143.800	19435.466
		68.249	953.980		
	$4_{14}-5_{24}$	5573.829	17936.007		
		74.062	935.259		
	$5_{15}-6_{25}$	5580.166	17915.641	5153.583	19398.571
		80.345	915.065	53.808	397.727
	$6_{16}-7_{26}$	5586.825	17894.286	5158.776	19379.045
		86.983	893.779		
	$7_{17}-8_{27}$	5593.662	17872.540	5165.122	19355.235
		93.779	872.039	65.350	354.383
$^PP_{2,N-2}$	$1_{10}-2_{20}$	5557.497	17988.715	5134.072	19472.290
		58.277	986.191	35.321	467.555
	$2_{11}-3_{21}$	5561.770	17974.897	5137.604	19458.905
		62.308	973.158	38.330	456.157
	$3_{12}-4_{22}$	5566.637	17959.180	5141.762	19443.168
				42.389	440.800
	$4_{13}-5_{23}$	5572.185	17941.300	5146.734	19424.385
		72.566	940.072	47.146	422.830
	$5_{14}-6_{24}$	5578.426	17921.226		
		78.778	920.096		
	$6_{15}-7_{25}$	5585.104	17899.798		
		85.515	898.481		

Table 8.19: Wavelengths and Frequencies of Transitions to $v'' = 1$: Π Subband

Branch	Rot. Trans.	L. Str. (Int.)*	(0,12,0) – (0,1,0)		(0,14,0) – (0,1,0)	
			λ_{air} (Å)	ν_{vac} (cm⁻¹)	λ_{air} (Å)	ν_{vac} (cm⁻¹)
$^r R_{0,N}$	$1_{10}-0_{00}$	3.00 (2.97)			5874.180	17018.935
	$2_{11}-1_{01}$	8.99 (7.63)	6422.778 21.861	15565.285 567.515	5871.259 71.619	17027.401 026.360
	$3_{12}-2_{02}$	5.69 (3.84)			5868.563 68.872	17035.226 034.328
	$4_{13}-3_{03}$	12.73 (6.30)			5865.957 66.258	17042.792 041.920
	$5_{14}-4_{04}$	6.41 (2.15)				
	$6_{15}-5_{05}$	11.91 (2.51)				
$^r Q_{0,N}$	$1_{11}-1_{10}$	8.99 (8.99)			5878.308 78.764	17006.984 005.665
	$2_{12}-2_{02}$	8.07 (7.05)			5879.814 80.047	17002.628 001.954
	$3_{13}-3_{03}$	24.73 (17.66)	6434.843 34.048	15536.102 538.021	5881.827 81.975	16996.809 996.383
	$4_{14}-4_{04}$	17.27 (9.41)	6437.445 36.637	15529.821 531.771	5884.132 84.236	16990.153 989.852
	$5_{15}-5_{05}$	44.74 (17.36)	6439.777 39.092	15524.198 525.850	5886.524 86.588	16983.248 983.064
	$6_{16}-6_{06}$	27.26 (7.00)			5888.830 88.900	16976.596 976.395
	$7_{17}-7_{07}$	63.48 (10.09)			5890.868 90.945	16970.725 970.504
$^p R_{2,N-2}$	$4_{13}-3_{21}$	4.33 (2.14)				
	$6_{15}-5_{23}$	13.37 (2.82)				
$^p Q_{2,N-1}$	$2_{11}-2_{21}$	4.95 (4.20)			5895.500 96.277	16957.390 955.156
	$4_{13}-4_{23}$	11.88 (5.88)			5897.937 98.422	16950.383 948.990
	$6_{15}-6_{25}$	17.64 (3.72)				

Table 8.20: Wavelengths and Frequencies of Transitions to $v'' = 1$: Π Subband

Branch	Rot. Trans.	L. Str. (Int.)*	(0,12,0) – (0,1,0) λ_{air} (Å)	ν_{vac} (cm⁻¹)	(0,14,0) – (0,1,0) λ_{air} (Å)	ν_{vac} (cm⁻¹)
$^PQ_{2,N-2}$	$3_{13}-3_{21}$	4.72 (3.37)			5899.560	16945.721
	$5_{15}-5_{23}$	2.53 (0.98)			5908.114	16921.187
$^PP_{2,N-1}$	$1_{11}-2_{21}$	8.90 (8.90)			5902.604 03.506	16936.982 934.395
	$2_{12}-3_{22}$	4.92 (4.30)			5907.937 08.508	16921.692^s 920.057
	$3_{13}-4_{23}$	11.00 (7.85)	6473.329 72.808	15443.734 444.977	5913.967 14.370	16904.439 903.289
	$4_{14}-5_{24}$	6.09 (3.32)	6481.055 80.543	15424.325 426.542	5920.562 20.896	16885.609 884.656
	$5_{15}-6_{25}$	13.30 (5.16)	6489.048 88.573	15406.324 407.452	5927.666 27.906	16865.374 864.691
	$6_{16}-7_{26}$	7.16 (1.84)			5935.129 35.349	16844.166 843.541
	$7_{17}-8_{27}$	15.17 (2.41)			5942.747 42.972	16822.574 821.939
$^PP_{2,N-2}$	$1_{10}-2_{20}$	4.81 (4.77)			5902.219	16938.087
	$2_{11}-3_{21}$	12.15 (10.32)	6465.508 65.049	15462.243 463.513	5907.012 07.695	16924.344 922.388
	$3_{12}-4_{22}$	7.88 (5.31)			5912.451 13.014	16908.774 907.164
	$4_{13}-5_{23}$	19.99 (9.90)			5918.635 19.106	16891.107 889.762
	$5_{14}-6_{24}$	12.17 (4.09)				
	$6_{15}-7_{25}$	28.42 (6.00)			5933.094 33.597	16849.943 848.516

Table 8.21: Wavelengths and Frequencies of Transitions to $v'' = 1$ and 3: Π Subband

Branch	Rot. Trans.	(0,16,0) - (0,1,0)		(0,14,0) - (0,3,0)	
		λ_{air} (Å)	ν_{vac} (cm^{-1})	λ_{air} (Å)	ν_{vac} (cm^{-1})
$^rR_{0,N}$	$1_{10}-0_{00}$			6662.809	15004.541
	$2_{11}-1_{01}$	5400.565	18511.436	6659.007	15013.108
		01.163	509.385	59.469	012.068
	$3_{12}-2_{02}$	5398.333	18519.087	6655.501	15021.017
		98.829	517.387	55.889	020.142
	$4_{13}-3_{03}$	5396.349	18525.896	6652.091	15028.718
		96.652	524.857	52.421	027.971
	$5_{14}-4_{04}$				
	$6_{15}-5_{05}$			6644.504	15045.877
				44.860	045.072
$^rQ_{0,N}$	$1_{11}-1_{10}$	5405.770	18493.612	6668.067	14992.710
				68.565	991.387
	$2_{12}-2_{02}$			6669.949	14988.710
				70.246	987.814
	$3_{13}-3_{03}$	5410.384	18477.841	6672.483	14982.787
				72.691	982.322
	$4_{14}-4_{04}$	5412.113	18471.937	6675.440	14976.150
				75.590	975.815
	$5_{15}-5_{05}$	5413.779	18466.252	6678.544	14969.191
		13.881	465.905	78.643	968.969
	$6_{16}-6_{06}$	5415.178	18461.481	6681.529	14962.503
		15.759	459.500	81.621	962.296
	$7_{17}-7_{07}$	5417.520	18453.501	6684.134	14956.673
		17.708	452.861	84.239	956.436
$^pR_{2,N-2}$	$4_{13}-3_{21}$				
	$6_{15}-5_{23}$				
$^pQ_{2,N-1}$	$2_{11}-2_{21}$	5421.070	18441.416	6698.680	14924.195
		22.024	438.172	6700.271	920.650
	$4_{13}-4_{23}$			6701.709	14917.450
				02.660	915.332
	$6_{15}-6_{25}$				

Table 8.22: Wavelengths and Frequencies of Transitions to $v'' = 1$ and 3: Π Subband Continued

Branch	Rot. Trans.	(0,16,0) − (0,1,0)		(0,14,0) − (0,3,0)	
		λ_{air} (A)	ν_{vac} (cm^{-1})	λ_{air} (A)	ν_{vac} (cm^{-1})
$^{P}Q_{2,N-2}$	$3_{13}-3_{21}$			6703.800	14912.796
				04.794	910.585
	$5_{15}-5_{23}$				
$^{P}P_{2,N-1}$	$1_{11}-2_{21}$	5426.306	18423.621	6707.859	14903.771
				09.578	899.954
	$2_{12}-3_{22}$				
	$3_{13}-4_{23}$	5437.770	18384.780	6722.405	14871.523
				23.241	869.675
	$4_{14}-5_{24}$	5443.105	18366.761		
	$5_{15}-6_{25}$	5448.562	18348.366	6740.086	14832.511
		48.848	347.404	40.620	831.336
	$6_{16}-7_{26}$	5454.304	18329.051		
		55.010	326.680		
	$7_{17}-8_{27}$	5461.388	18305.278	6759.774	14789.313
		61.688	304.273	60.151	788.486
$^{P}P_{2,N-2}$	$1_{10}-2_{20}$				
	$2_{11}-3_{21}$	5430.790	18408.411	6713.421	14891.425
		31.662	405.454	14.707	888.573
	$3_{12}-4_{22}$	5435.404	18392.785	‥	
		36.135	390.312		
	$4_{13}-5_{23}$	5440.888	18374.248	6727.883	14859.414
		41.380	372.586	28.750	857.500
	$5_{14}-6_{24}$				
	$6_{15}-7_{25}$			6745.930	14819.662
				46.754	817.852

Table 8.23: Wavelengths and Frequencies of Transitions to $v'' = 0$ and 1: Φ Subband

Branch	Rotational transition	Line strength (intensity)*	(0,12,0)−(0,0,0) λair (Å)	νvac (cm⁻¹)	(0,14,0)−(0,0,0) λair (Å)	νvac (cm⁻¹)	(0,14,0)−(0,1,0) λair (Å)	νvac (cm⁻¹)
$^rR_{2,N-1}$	$3_{31}-2_{21}$	14.84 (23.64)	6065.329	16 482.587	5575.496	17 930.644	5922.523	16 880.018
					76.961	925.934	24.360	874.783
	$4_{32}-3_{22}$	7.76 (9.22)			5574.240	17 934.683		
					75.240	931.469		
	$5_{33}-4_{23}$	16.19 (13.34)	6061.513	16 492.964	5573.330	17 937.612	5920.053	16 887.063
			62.695	489.750	74.062	935.259	20.896	884.656
$^rR_{2,N-2}$	$4_{31}-3_{21}$	15.11 (17.94)	6063.701	16 487.015	5574.552	17 933.680	5921.442	16 883.101
			65.184	482.983	75.565	930.425	22.649	879.660
$^rQ_{2,N-1}$	$4_{31}-4_{23}$	9.77 (11.60)	6078.936	16 445.695	5587.430	17 892.348	5935.932	16 841.889
			80.414	441.696	88.424	889.165	37.102	838.568
$^rQ_{2,N-2}$	$3_{31}-3_{21}$	5.04 (8.03)			5585.829	17 897.478	5934.151	16 846.943
					87.233	892.979	35.858	842.099
	$5_{33}-5_{23}$	12.68 (10.45)	6083.529	16 433.277	5591.946	17 877.899	5940.893	16 827.826
					92.666	875.597	41.742	825.421
$^pQ_{4,N-3}$	$4_{31}-4_{41}$	4.68 (5.56)			5627.035	17 766.418	5987.041	16 698.116
					28.461	61.916	88.885	692.976
$^pQ_{4,N-4}$	$5_{33}-5_{41}$	7.83 (6.45)	6128.329	16 313.146	5629.753	17 757.839		
					30.867	754.326		
$^pP_{4,N-3}$	$3_{31}-4_{41}$	20.44 (32.56)	6139.942	16 282.294	5638.483	17 730.346	6000.027	16 661.976
					40.311	724.599	2.315	655.627
	$5_{33}-6_{43}$	22.76 (18.75)	6152.748	16 248.403	5650.366	17 693.057	6013.397	16 624.931
			54.323	244.245	51.402	689.816	14.655	621.453
$^pP_{4,N-4}$	$4_{31}-5_{41}$	21.37 (25.37)	6146.116	16 265.935	5644.155	17 712.528	6006.417	16 664.251
			48.031	260.869	45.509	708.281	8.115	639.547

*Intensities calculated for $T = 180\,\mathrm{K}$ and rotational levels of $\tilde{A}$ (0,15,0) relative to 0_{00}.

8.3 Bibliography

[1] H. Lew. Electronic spectrum of H_2O^+. *Canadian Journal of Physics*, 54(20):2028–2049, 1976.

[2] H. Lew and R. Groleau. Electronic spectrum of D_2O^+. *Canadian Journal of Physics*, 65(7):739–752, 1987.

Index

Author index

Printed and bound by CPI Group (UK) Ltd, Croydon, CR0 4YY

24/10/2024

01778301-0010